Distinguished Dissertations

Springer
London
Berlin
Heidelberg
New York
Barcelona
Hong Kong
Milan
Paris
Santa Clara
Singapore
Tokyo

Other titles published in this Series:

Extensional Constructs in Intensional Type Theory
Martin Hoffman

Search and Planning Under Incomplete Information: A Study Using Bridge Card Play
Ian Frank

Theorem Proving with the Real Numbers
John Harrison

Games and Full Abstraction for a Functional Metalanguage with Recursive Types
Guy McCusker

Hardware Evolution: Automatic Design of Electronic Circuits in Reconfigurable Hardware by Artificial Evolution
Adrian Thompson

Models of Sharing Graphs: A Categorical Semantics of **let** *and* **letrec**
Masahito Hasegawa

Chris Greenhalgh

Large Scale Collaborative Virtual Environments

Springer

Chris Greenhalgh, BA, PhD
Faculty of Science, School of Computer Science and Information Technology,
University Park, University of Nottingham, Nottingham, NG7 2RD, UK

Series Editor

Professor C.J. van Rijsbergen
Department of Computing Science, University of Glasgow, G12 8RZ, UK

ISBN 978-1-4471-1222-8 Springer-Verlag London Berlin Heidelberg

British Library Cataloguing in Publication Data
Greenhalgh, Chris
Large scale collaborative virtual environments. -
(Distinguished dissertations)
1.Virtual reality 2.Computer network architectures
3.Distinguished operating systems (Computers) 4.Multimedia
systems
I.Title
006
ISBN-13:978-1-4471-1222-8 e-ISBN-13:978-1-4471-0867-2
DOI: 10.1007/978-1-4471-0867-2

Library of Congress Cataloging-in-Publication Data
A catalog record for this book is available from the Library of Congress

Typesetting: Camera-ready by author

34/3830-543210 Printed on acid-free paper

To Vanessa, Susannah and Timothy.

Preface

Collaborative virtual environments (CVEs) are multi-user virtual realities which seek to actively support communication and cooperation between distributed users.These 3D audio-graphical environments have the potential to:

- scale to very large numbers of simultaneous participants much more effectively than other forms of (non-spatial) computer-mediated communication such as video or audio conferencing;
- provide a unified context for communication and information, for example allowing an architectural team to have a tele-conference within a visualisation of a planned building;
- maintain autonomy of individuals, and an appropriate balance of power and influence between participants;
- support "natural" uses of space in communication, such as dynamic group formation, gesture and gaze direction; and
- allow users to easily maintain peripheral awareness of nearby activities, and to assess the states of systems and activities at a glance.

When we seek to meet these objectives we are faced with many technical and conceptual challenges. For the systems which do exist, we want to know how effective they are and, indeed, whether our objectives are actually achievable.

Origins

I have been researching, implementing, using and deploying collaborative virtual environments for just over five years, first as a post-graduate student, and more recently as a lecturer, both at the University of Nottingham. This book is my doctoral thesis, which was one of two winners of the 1998 Distinguished Dissertations in Computer Science competition. This competition is run by the Conference of Professors and Heads of Computing (CPHC), in conjunction with the British Computer Society (BCS), in order to select for publication a few of the best British PhD/DPhil dissertations in computer science.

My thesis is presented here complete and unchanged, except for reformatting and the removal of two appendices which are not relevant for publication in this form (colour images and video presentation contents). I have also added an index to help with selective reading and reference use. It is a great pleasure to see this in print, and I hope that you will find it interesting and gain insight from it.

Intended Audience

There are four intended audiences for this book. First, (because of its origins as a PhD dissertation) it is directly relevant to other researchers and post-graduate students working in the area of multi-user virtual reality. It is particularly pertinent for researchers concerned with:

- architectures for distributed virtual reality systems;
- network requirements of these systems;
- use and user evaluation of collaborative virtual environments; and
- models of interaction and communication in space-based multi-user systems.

Second, it can be used by researchers in related areas, such as single-user virtual reality, computer supported cooperative work and distributed systems, to gain a detailed understanding of the issues peculiar to systems and environments of this kind.

Third, it should be accessible to any advanced computer-literate reader wishing to explore the current state of the art in specifically multi-user virtual reality.

Fourth, it may serve as a model for future students (one of the objectives of the Distinguished Dissertations series) since it has been judged to be among the best current dissertations in the area of computer science. The competition also requires that it combine a noteworthy contribution to the subject with a high standard of exposition, in such a way that it is accessible to those with other specialities.

Content and Structure

This book is based around two prototype collaborative virtual environment systems, MASSIVE-1 and MASSIVE-2, which I have designed and implemented. These systems realise and extend a spatial model of interaction, which provides a conceptual and computational basis for managing communication and interaction. I have sought to combine two major perspectives:

- the user's perspective within the everyday social world - considering the kinds of interaction and control which are afforded to a participant in a CVE; and
- the network's perspective - considering the form and requirements of the communication which facilitates this interaction.

Section 1.4 (page 13) explains the structure of the book and gives suggestions for reading order. The overall form (from page 15) is shown below.

Chapter 1. Introduction

Chapter 2. CVEs: a review

Key concept	Theory	Prototype overview	Social issues	Communication issues
Part I: Direct relationships	Chapter 3: Original spatial model	Chapter 4: MASSIVE-1 overview	Chapter 5: Awareness relationships	Chapter 6: Spatial Trading
Part II: Indirect relationships	Chapter 7: Third party objects	Chapter 8: MASSIVE-2 overview	Chapter 9: Regions and awareness	Chapter 10: Awareness driven communication

Chapter 11. Summary, conclusions and future work

Part I deals with the concepts, design, user-oriented evaluation and network requirements of MASSIVE-1, which is a unicast-based CVE implementing the spatial model of interaction of Benford and Fahlén (reviewed in chapter 3).

Part II does the same for MASSIVE-2, which is a multicast-based CVE implementing the proposed "third party object" extension to the spatial model. (See section 2.2.2 for a comparison of unicast and multicast communication.)

Subsequent Developments

I completed this dissertation in September, 1997. Since then work has continued on MASSIVE-2 and its successor, however this cannot be integrated here because the thesis is published "as is".

Perhaps the single most significant event of the past year has been the use of MASSIVE-2 for staging "Out Of This World", an experiment in Inhabited Television (see section 11.3.3). Out Of This World can be viewed as a logical extension of the Now96 event (described in section 9.2.1). Eight inhabitants (members of the public sitting at PCs wearing microphone-headsets) formed two teams that competed in a series of co-operative games, the winners gaining the right to escape from the doomed world. The show was projected live in a nearby auditorium to the show's viewers (also members of the public). Each team was led by a captain, played by a paid performer, who was immersed in the virtual world using a headset and tracking technology. The games were overseen by a host, who appeared in the world as a video texture. Whilst each performance took place, four virtual camera operators filmed the action. An experienced television director mixed the final show using standard outside broadcast equipment.

Out Of This World was a great technical success, and the performances (four in total) were coherent and accessible. The inhabitants were generally involved and able to interact and collaborate. We believe that the event demonstrated the feasibility and potential effectiveness of Inhabited Television. The next key areas for development are the creation of engaging content for this format, its integration with longer-term communities of participants (at all levels - authors, performers, inhabitants and viewers) and the continuing quest to practically support much larger numbers of simultaneous participants.

Acknowledgements

First and foremost, I would like to thank my supervisor, Steve Benford, for providing an excellent environment in which to work and a firm conceptual and ideological foundation on which to build. Our past and present collaboration has been invaluable and most enjoyable.

I would also like to thank all those members of the Communication Research Group, past and present, who have shared their time, friendship and interests. Dave Snowdon made many valuable contributions, including discussion and implementation. Adrian Bullock was the linchpin that made trials happen and machines work. Maarten Stenius (a visiting researcher from SICS) ported WWW-3D to MASSIVE-2. Many thanks also to Karsten Bormann, Rob Clarke, Neil Cook, Mike Craven, Mike Fraser, Tony Glover, Rob Ingram, Boriana Koleva, David Lloyd, Jim Purbrick, Mina Radenkovic, Gail Reynard, Marcus Roberts, Ian Taylor, Jolanda Tromp and Ivan Vaghi.

Finally, I would like to thank Jon Crowcroft, my external examiner, for supporting my submission to the Distinguished Dissertations in Computer Science competition, and the panel members and other reviewers for their favourable consideration of my work.

Abstract

This thesis is concerned with the theory, design, realisation and evaluation of large-scale collaborative virtual environments. These are 3D audio-graphical computer generated environments which actively support collaboration between potentially large numbers of distributed users. The approach taken in this thesis reflects both the sociology of interpersonal communication and the management of communication in distributed systems.

The first part of this thesis presents and evaluates MASSIVE-1, a virtual reality tele-conferencing system which implements the spatial model of interaction of Benford and Fahlén. The evaluation of MASSIVE-1 has two components: a user-oriented evaluation of the system's facilities and the underlying awareness model; and a network-oriented evaluation and modelling of the communication requirements of the system with varying numbers of users.

This thesis proposes the "third party object" concept as an extension to the spatial model of interaction. Third party objects can be used to represent the influence of context or environment on interaction and awareness, for example, the effects of boundaries, rooms and crowds. Third party objects can also be used to introduce and manage dynamic aggregates or abstractions within the environments (for example abstract overviews of distant crowds of participants). The third party object concept is prototyped in a second system, MASSIVE-2.

MASSIVE-2 is also evaluated in two stages. The first is a user-oriented reflection on the capabilities and effectiveness of the third party concept as realised in the system. The second stage of the evaluation develops a predictive model of total and per-participant network bandwidth requirements for systems of this kind. This is used to analyse a number of design decisions relating to this type of system, including the use of multicasting and the form of communication management adopted.

Contents

1.
Introduction

Collaborative virtual environments (CVEs) are computer-based systems which actively support human collaboration and communication within a computer based context. The particular focus of this thesis is on the theory, realisation and requirements of large scale collaborative virtual environments, i.e. CVEs which will support large numbers of simultaneous users. The particular forms of CVE considered in this thesis are three dimensional audio-graphical virtual environments - distributed virtual reality for collaboration. Section 1.1 introduces CVEs and the motivations for exploring and developing this technology. Section 1.2 describes the general issues of scale which are significant for CVEs and identifies those which are key with respect to this work. Section 1.3 describes the particular style and approach adopted in this thesis which is based on the concept of "awareness". Finally, section 1.4 describes the structure of this thesis as a whole and includes guidance on reading order and selection of material. Before these, however, some more general issues must be covered which relate to this thesis as a whole.

The work presented here falls firmly within the boundaries of computer science however many of the motivations which underlie it flow from work in the field of social studies. Concerns from both of these disciplines are represented in this thesis. On the one hand there is a perspective looking "down" from the everyday social world and considering the kinds of interaction and control which are afforded to users of CVEs. On the other hand there is a perspective which looks "up" from the world of computers and networks and considers the form and requirements of the distributed systems which facilitate this interaction. A fundamental argument of this thesis is that these two perspectives must be complementary and interlocking if computer based systems are to approach the richness of our everyday - social - existence. This thesis explores and extends the so-called "spatial model of interaction" [Benford and Fahlén, 1993] which is proposed as a suitable framework within which these two perspectives might be integrated.

The spatial model of interaction was previously proposed and developed by Benford, Fahlén and others; this prior work is reviewed in chapter 3. The work presented in this thesis comprises a fundamental extension to the spatial model of interaction which is bracketed (both in time and in presentation) between two major and distinct phases of prototyping and evaluation. This two-stage evolution has been preserved in the organisation of this thesis (refer to section 1.4 for details of thesis structure).

The principle contributions of this thesis are:

- The MASSIVE-1 and MASSIVE-2 systems as a basis for experimentation and development (described in chapters 4 and 8, respectively).
- A user-oriented evaluation of the spatial model of interaction as implemented in MASSIVE-1, with recommendations for future models and systems (chapter 5).
- A new distributed system service concept of "spatial trading", which extends the concept of attribute-based trading (as in ODP [ITU-T, 1995] for example) in line with the principles of CVEs (chapter 6).
- The third party object concept, which extends the spatial model of interaction to include the effects of context or environment on interaction (chapter 7). This is prototyped in MASSIVE-2 (chapter 8) and evaluated from a user perspective (chapter 9).
- A flexible and dynamic multicast-based distribution approach for CVEs which is based on third party objects and prototyped in MASSIVE-2 (chapter 10).
- A predictive model of network bandwidth requirements (total and per-participant) for peer-to-peer CVEs such as MASSIVE-1 and MASSIVE-2 which is used to analyse a number of design aspects of such systems (chapter 10).

Much of the work presented in this thesis has already been presented at conferences or published in journals (e.g. ACM CHI'97 [Benford, Greenhalgh and Lloyd, 1997], IEEE ICDCS'95 [Greenhalgh and Benford, 1995c], ACM ToCHI [Greenhalgh and Benford, 1995a], ECSCW'95 [Greenhalgh and Benford, 1995c], Presence [Benford et al., 1995] - see appendix B for a complete list of related publications). In some cases these papers can provide additional detail and examples which have been omitted from this thesis for the sake of focus and (relative) brevity.

At the outset the author wishes to recognise the invaluable role played by Dr S.D. Benford in supervising the work presented here. The key concepts of the spatial model extensions presented here are the author's own original work, but naturally they have been refined and clarified in many valuable discussions with Dr Benford and other colleagues. Dr Benford and others have also continued to develop and extend the applicability of these concepts beyond that presented here. The author also wishes to recognise the valuable contributions made by Dr D.N. Snowdon who has been closely involved with parts of this work, in particular contributing the generic graphics and device handling components of the second prototype system.

Having dealt with these general issues, the next section introduces the idea and practice of collaborative virtual environments.

1.1. Collaborative Virtual Environments

This section seeks to answer two questions: what is a collaborative virtual environment and why are they interesting? These will be addressed in turn. So, first, what is a collaborative virtual environment?

There are a number of characteristics which are common to collaborative virtual environments:

- they are multi-user computer-based systems which support geographically dispersed users;
- users are able to communicate and collaborate, often in a number of different ways;
- there is a notional space or world - the virtual environment - in which this activity is situated;
- each user is explicitly represented or "embodied" within the virtual environment and is made visible (and audible, etc.) to other users by means of this embodiment; and
- each user is autonomous and able to move about independently within the virtual environment.

Note that all collaborative systems support multiple users but not all multi-user systems support collaboration. For example, it is noted by Rodden et al. [1992] that one of the functions of a database system is to isolate its users so that each one can work as if they were the sole user of the database. Such a system supports multiple users but provides no meaningful support for collaboration - it provides no information about other users or their current activities and no opportunity for communication. Supporting collaboration has its own characteristic requirements. Much of this thesis (and the spatial model of interaction) is concerned with improving the support provided for collaboration based on observations of critical issues in everyday interactions in the physical world drawn from the discipline of social studies.

While sharing a common philosophy collaborative virtual environments can be realised using a wide range of presentation and distribution technologies. The form of presentation can vary from plain text through windowed 2D graphics to fully immersive virtual reality. The form of distribution can be none (e.g. a single process accessed via telnet) or one or more of a wide range of network architectures based on different execution and communication models. The range of possible instantiations of a CVE runs from typical MUDs (Multi-User Dimensions/Dungeons) at one extreme to multi-user immersive virtual reality at the other. Consider each of these in turn, starting with MUDs. Users access a MUD using a simple terminal program and all interaction is by means of text messages. Multiple simultaneous users can "see" each other (i.e. read text descriptions of other users), "talk" to each other (i.e. type text messages which others read) and "walk" about within the MUD's electronic world. A single server process typically handles all processing. On the other hand, access to an immersive multi-user virtual environment requires a great deal of specialised equipment. Users may wear head mounted displays, have their body movements, gestures, facial expressions and gaze direction captured in electronic form, they may receive tactile feedback and have real-time audio and video communication facilities. So users of an immersive multi-user virtual environment can see expressive articulated representations of other users, can speak to each other directly, use normal gestures and expressions in conversation and may be able to move about in the virtual world by walking in the real world. In such systems processing will typically be distributed between a number of computers and high performance networks may be required to link the various systems.

The types of collaborative virtual environment being considered in this thesis are primarily three dimensional (although the concepts of the spatial model can be applied in other kinds of spaces such as graphs - see for example [Rodden, 1996]). The systems considered in this thesis also provide relatively rich interaction including 3D graphics, real-time audio, dynamic maps and simple text messages. As an example, figure 1.1 on page 4 shows an image from the MASSIVE-1 system which is the first of the prototype CVE systems described in this thesis (in chapters 4 through 6). In the foreground is a recently arrived user. They are approaching a group of participants who are waiting in the foyer world for a meeting to start. Each user is represented within the virtual world by a graphical embodiment. Each user can move independently, and the position and orientation of each user's embodiment provides

Fig.1.1: a small group of participants in the MASSIVE-1 CVE

clues about what they are seeing and doing. Users can communicate using simple gestures, real-time audio and also text messages (via another interface which is not shown here). The arches in the background are gateways to other virtual worlds. Between the two gateways is a message board which allows semi-persistent text messages to be left in the virtual world.

For further material about CVEs and multi-user virtual reality systems the reader is referred to the literature review in chapter 2 and also to chapters 4 and 8 which describe the two prototype CVE systems which were created in the course of the work presented in this thesis.

Having introduced collaborative virtual environments it is now time to consider the second question: why are CVEs interesting? The answer is presented in two parts, first from the perspective of collaboration and second from the perspective of computer networking. In terms of support for collaboration CVEs are interesting for a number of reasons, some of which are described below.

- Supporting "natural" spatial communication. Space has a social significance which is important for real-world interaction. Firstly, significant elements of communication such as gaze direction and gesture depend upon a spatial reference frame. And more generally, space can also be viewed as a resource for managing activity and interaction (see [Benford et al., 1994] which builds on [Giddens, 1984]). Everyday examples of the use of space as a resource can be seen in the fluid formation of conversational groups in informal social gatherings. Spatial factors such as relative positions and orientation of participants, posture and speed of movement consciously and unconsciously convey information such as availability for conversation, interest and intended actions. Shared space is a fundamental concept in CVEs which, it is hoped, can be used to support some of these everyday functions and affordances of the social use of space.
- Supporting peripheral awareness. Studies of real-world activities in London Underground control rooms [Heath and Luff, 1991] and in air traffic control [Hughes et al., 1992] observe that co-located workers maintain and make use of an ongoing awareness of the activities of others within their environment, even when not explicitly cooperating. The situation of work within a shared context is a powerful method for achieving this kind of peripheral awareness.
- Unifying communication and information. A great deal of work is being done in the area of visualisation - creating (often graphical) representations of information to enhance access and comprehension. Such visualisations may be situated within a collaborative virtual environment so that users have combined access to information alongside facilities for communication and cooperation. Such environments, with users virtually co-located with the information with which they are working, are also known as a Populated Information Terrains (PITs) (see for example [Benford et al., 1994]).
- Maintaining autonomy. With a CVE-like approach each user retains their own perspective and independence of movement and activity. This is in contrast to, for example, the pure WYSIWIS (What You See Is What I See) approach to multi-user 2D window-based collaboration support (c.f. [Stefik et al., 1987]) in which

each collaborator sees exactly the same view and is correspondingly more tightly bound to a single activity. CVEs have the potential to support spontaneous encounters, loose and informal collaborations, browsing and exploration which are potentially problematic for more formal or prescriptive approaches to collaboration support such as work flow systems (see for example [Bowers, 1995]).

- Scaling to large numbers of participants. The final advantage of CVEs from the perspective of collaboration is that there is a clear potential to support extremely large number of simultaneous participants. In the physical world effects such as perspective, occlusion and dispersion allow us to be aware of and to interact with a relatively large, continuous and consistent subset of the total environment. Applying the same kinds of notions in CVEs would allow large numbers of simultaneous participants to communicate and interact with varying (and individually appropriate) levels of fidelity. It is very hard to see how other real-time collaboration technologies such as video and audio conferencing could cater for very large numbers of mutually aware participants.

In terms of computing, and in particular computer networking, CVEs have a number of other interesting (and challenging) aspects which motivate their exploration and development. These are listed below.

- Mixed traffic types. CVEs provide a natural context in which to combine distributed data services with audio and video flows, all in the context of real-time interaction with geographically dispersed users. For each information flow different levels of fidelity may be required by different observers at the same time and by the same observer at different times (e.g. depending on proximity within the virtual world).
- Group participation. CVEs aim to support collaboration between arbitrary groups of people. This raises issues concerning the appropriate representation of and support for these communicating groups within the computational and networking domains.
- Dynamic connection and membership. It is clear from the sociological motivations behind CVEs that users will be continually moving within and between virtual worlds, as well as entering and leaving the system. Consequently the communication requirements are highly dynamic.
- Dynamic negotiation of quality of service (QoS). One note-worthy aspect of these dynamic communication requirements is that network quality of service will be subject to continuous and ongoing re-negotiation as users move between worlds, regions and collaborators.
- Reliability and consistency. There remain open issues about appropriate choices and technologies for achieving a suitable balance between notional correctness and timeliness in CVEs (e.g. trading degrees of reliable delivery and ordering of messages against speed of interaction in and with the virtual world).
- Scale. Finally, as noted above, CVEs have the potential to involve very large numbers of simultaneous participants. For example, it is proposed that US DOD virtual military exercises might involve thousands or even hundreds of thousands

of real or simulated simultaneous participants [Macedonia et al., 1994]. On an even larger scale, one can imagine delivering CVE technologies to domestic consumers as a kind of "inhabited television"; if this ever became a reality then the potential number of simultaneous participants could be many millions.

This section has introduced the key characteristics of CVEs. It has also presented motivations for the study and development of CVEs from the twin perspectives of collaboration and distributed systems. It is suggested that CVEs are both an appropriate technology for supporting collaboration and an interesting class of application to drive the continuing evolution of computer communication services. The next section explores in more detail the computational and networking issues which relate to the scalability of CVEs, i.e. to the ability to support large and increasing numbers of simultaneous users.

1.2. Issues of Scale

There are a number of aspects of CVEs which may be described in terms of "scale" (i.e. the size or dimension of the problem or system). This section identifies the particular aspects and interpretations of scale which are critical in this thesis. It then considers the factors which potentially limit the scale of collaborative virtual environments and identifies the particular approaches to scale adopted in this work. First, some of the areas in which scale may be considered are listed below.

- Geographical distance between participants, e.g. co-located or remote.
- Network "distance" between participants, e.g. latency, jitter, packet loss.
- Total user population, i.e. all those who use the system, whether they are currently doing so or not.
- Number of simultaneous participants.
- Scope of participant awareness, i.e. the fraction of the total environment to which a single participant has access at any moment.
- Complexity of the virtual environment.
- Richness of communication, e.g. number of media, level of detail or fidelity.
- Variability of delivery platforms, e.g. different machine capabilities, different peripherals, different network facilities and technologies.

Scalability refers to the effects of increasing the scale of a problem or system. So a scalable system is one for which the cost of increases in scale is regarded as "small" or "acceptable". One way of representing this is in terms of the order of the cost with respect to the dimension being considered. Non-polynomial orders would normally be considered extremely non-scalable. Scale $O(n)$ and below (and $O(n \cdot \ln n)$ for pair-wise processes) would typically be considered scalable (few things in life - or computer science - are free).

The qualification "large scale" in the title of this thesis refers to a number of choices in terms of the dimensions listed above, but primarily it refers to scalability with respect to the number of simultaneous participants. The reason for this focus and the other choices which have been made are described below.

- Geographical distance between participants: participants are assumed to be remotely located at the level of the current Internet, i.e. dispersed over sites, countries and continents.
- Network distance: wide area use is assumed, which follows from geographical distance. This is essentially a background constraint which has influenced the internal construction of the prototype systems (e.g. by making few assumptions concerning delay or jitter) but which is not a focus of the theory or evaluation work.
- Total user population: this dimension is not explicitly considered. The total user population has costs in terms of persistent storage requirements, management and billing but these are outside the primary real-time focus of this thesis. In the context of real-time interaction the number of *simultaneous* participants is much more significant than the total user population (any number of whom may not be participating in the system at any given time).
- Number of simultaneous participants: variation of this dimension is a central focus of the computer networking aspects of this thesis - this is developed in the later parts of this section.
- Scope of participant awareness: this thesis deals explicitly with the issues of choosing which subset of the environment to interact with. However, the focus of this work is less concerned with *increasing* this scope of awareness as it is with *choosing* it appropriately (see the discussion which follows on computational limitations).
- Complexity of the virtual environment: the general focus of this thesis is on human-human interaction - computer mediated communication rather than data visualisation. As such, the complexity of the environment itself is treated as a secondary concern. Since CVEs exist (one hopes) for the ultimate benefit of human users this seems to be a good starting point.
- Richness of communication: this thesis assumes a certain minimum level of potential communication to support collaboration. This includes 3D graphics and real-time audio. While other media (excepting text) are not dealt with explicitly, the spatial model of interaction can, in principle, accommodate arbitrary media.
- Variability of delivery platforms: catering to the variability of target platforms is an integral part of the approach presented here. The first prototype system is particularly interesting as an example of support for diverse delivery platforms. This issue also informs the design of awareness scoping, mentioned above.

The focus of CVEs as presented here is on supporting real-time interaction. Some work is being done on integrating asynchronous interaction but this is not considered here. Consequently, as noted above, the number of simultaneous users is much more significant for general system cost and complexity than is the total user population.

When considering large and increasing numbers of simultaneous participants it becomes apparent that there are a number of potential bottlenecks or critical factors to consider. These critical factors are illustrated in figure 1.2 on page 9 and described below. The descriptions below also identify the factors which are key with respect to the work presented in this thesis.

Fig.1.2: critical points for scalability with simultaneous use

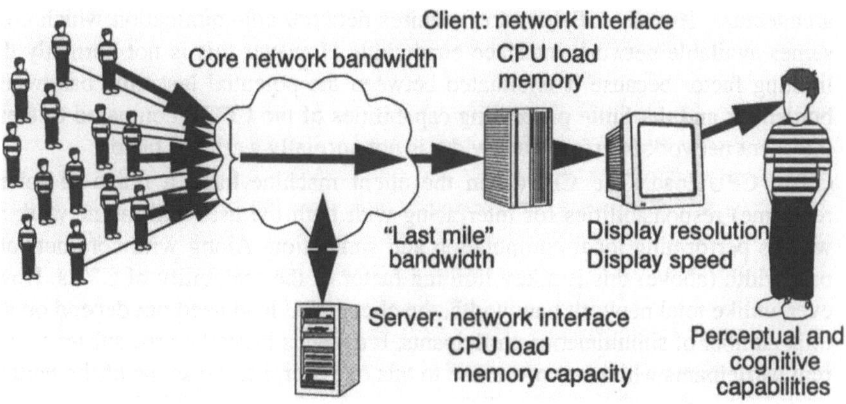

- Server: network interface, CPU load and memory capacity. Most CVE systems will include components which provide shared services of some form, although they may not be accessed in a traditional client-server style. These services may include trading and brokering, provision of world content and handling of persistence of worlds. Increasing the number of simultaneous users will impose a direct cost on these server components which may be measured in terms of increased communication, processing and storage requirements. However, the focus of this thesis with respect to scalability is on the delivery path to each participant, rather than the central services. This is partly because of the emphasis on supporting real-time human-human communication and partly because of the selection of peer-to-peer and multicast communication technologies for the prototype systems.
- Core network bandwidth. As the number of geographically dispersed participants increases the requirements for core network bandwidth must also increase. Core network bandwidth is a key limitation on the potential number of simultaneous users, and is one of the primary measures used in the network evaluations in chapters 6 and 10.
- "Last mile" bandwidth. In some situations (e.g. domestic access) the user access network can be particularly limited. This is the section of the network which carries information "the last mile" from the core network provider to the home user. Use of modems over dial-up connections is a clear example of this. This is a key limitation to consider for this market which tends to mitigate in favour of a cli-

ent-server system. However, in this thesis the last mile bandwidth has been given less emphasis on the basis that emerging subscriber loop technologies such as cable modems and ADSL over normal telephone wiring have the potential to dramatically increase access bandwidth over the next two to five years. More weight has been given to the client machine limitations described below.

- Client network interface. Typically for a CVE each participant is using their own machine to access the CVE. This machine is referred to here as a "client" machine or the user's machine, but without implying the use of a client-server architecture. Human collaboration requires network communication which consumes available network interface bandwidth. However this is not normally the limiting factor because it is situated between the potential last mile bandwidth bottleneck and the finite processing capabilities of the CPU - compared to these the client network interface bandwidth is not normally a critical factor.
- Client CPU load. The CPU(s) in the client machine have a range of (often real-time) responsibilities for interfacing with both the user and the network as well as performing local computation and simulation. Along with core network bandwidth (above) this is a key limiting factor of the scalability of CVEs. However, unlike total network bandwidth, the client CPU load need not depend on the total number of simultaneous participants. It depends instead on the subset of current participants which are presented to this one user, i.e. the scope of the participant's awareness. This is one of the main concepts which underlies this thesis. One application of the spatial model of interaction and the third party object extension proposed in this thesis is as a mechanism for dynamically determining the appropriate scope of participant awareness. The potential impact of this on client machines is considered especially in chapters 6 and 10.
- Client memory capacity. For extremely complex virtual environments (e.g. city walk-through) this is likely to become a major limitation. However with the focus in this thesis on interaction and continuous media (e.g. audio support) and the recent increases in typical workstation memory capacity this is not a key issue in this work. It is indirectly addressed by the same mechanisms which mitigate CPU load (above).
- Display resolution. A visual display (or audio display, etc.) can only present a finite resolution and finite breadth of content. This is an important consideration in some application areas (e.g. medial imaging) and also when dealing with significant amounts of text. As noted in section 5, a more significant limitation with regard to communication and collaboration can be the limited field of view available with desktop or typical head-mounted displays. While being noted, this issue is not a key focus of this thesis.
- Display speed. This work has not focused on raw interaction speed (e.g. the normal mode of delivery is via desktop workstations with mouse and keyboard input). A certain minimum responsiveness is required but display speed has not been a key consideration.
- Perceptual and cognitive capabilities. It is essential to remember that a person (or group of people) is the ultimate participant in a CVE. They have their own innate capabilities and limitations. For example, few people can fully participate in more

than one simultaneous conversation - other nearby activities and conversations receive a different degree of attention. People cannot perceive arbitrarily quiet sounds, see through opaque objects or perceive arbitrarily small or distant objects. This is "good news" for the scalability of CVEs: a system may have one million simultaneous users but any single user will only need to be presented with a fraction of that information (voices, faces, gestures, etc.). Interaction in the virtual environment may actually be significantly *more* limited than in the physical world, but the same kind of consistent constraints might make this restricted awareness understandable and controllable for normal users. This is fundamental to the use of the spatial model of interaction to facilitate large-scale collaborative virtual environments (which is the subject of this thesis).

This introduction to the areas of scalability in collaborative virtual environments concludes with a summary of the key issues of scale with respect to this thesis.

- The primary variable is the number of simultaneous participants.
- The two key technical limitations considered are core network bandwidth and the load imposed on a single user's machine.
- With respect to scale, the key principle or approach is to exploit (literally and metaphorically) limitations in people's perceptual and cognitive capabilities.

The goal is to provide an *adequate* and socially and psychologically *appropriate* basis for collaboration, rather than complete knowledge or perfect fidelity. The next section of this chapter describes the particular approach which is adopted in this thesis to combining and reconciling consideration for social issues with concerns for scalability and implementation.

1.3. Approach

The final area addressed in this introductory chapter concerns the particular style and approach adopted in this thesis. This is based on the concept of "awareness". This is a key concept which has emerged from the field of Computer Supported Cooperative Work and is inspired by the observation and analysis of both "everyday" and computer-mediated collaboration (see chapter 3 for more consideration of awareness). Awareness is probably the most important concept in the spatial model of interaction which is the basis of much of the work presented here. But awareness also represents a very general idea or approach to communication and collaboration. In this thesis, as in the spatial model of interaction, awareness quantifies the degree, nature or quality of interaction between two objects. It has a strong social dimension which reflects the way in which everyday interaction is performed and managed (see section 3.1). Awareness may also be represented or embodied within a working system as a computational model which controls and manages other aspects of system behaviour and activity such as presentation and quality of service negotiation. Figure 1.3 on page 12 shows the "cycle of awareness" as it is considered in this thesis.

Fig.1.3: the cycle of awareness

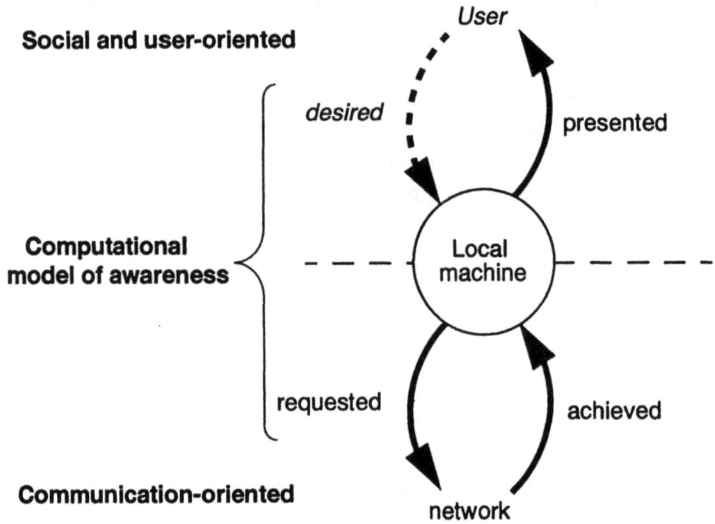

At the top of this awareness cycle is the user, or, more accurately, the richly connected social world of which the user is a part. Each user has their own particular goals and requirements and these imply a particular "pattern" of desired awareness, i.e. what (and who) they want to be aware of. These requirements must be captured or estimated in some way by a CVE system; this is the first role of a computational model of awareness. Depending on this (approximate) *desired* awareness, the user's local machine will filter the information which is *presented* to them. This awareness loop between a user and their machine determines what the user can express and how they can control their interaction with and through their local computer. This interaction is dealt with in this thesis under the heading of "social" or "user-related" issues, and is a particular emphasis of chapters 5 and 9.

The bottom of the awareness cycle is the connection between a user's "own" machine and the rest of the computational world, i.e. the network and the other machines on it. Based on the user's inferred desires the local machine determines which remote resources and flows of information the user requires access to. It then *requests* and negotiates access to these remote resources. The limitations of the network and of the various computational devices involved will mean that the information which is actually obtained may only approximate that which was requested. This *achieved* awareness is then available to the local machine to be filtered and presented to the user as before, completing the cycle of awareness. System scalability will be determined to a large extent by the interactions in this communication-oriented part of the cycle. This is dealt with in this thesis under the heading of "communication issues", and is a particular emphasis of chapters 6 and 10.

The next and final section of this chapter explains the structure of this thesis in relation to the general approach described here.

1.4. Thesis Structure and Reading Plan

This section describes and explains the structure of this thesis and offers some suggestions for reading and understanding it (for example, for selective reading). The concept of awareness, as introduced in the previous section, is the key concept around which the chapters are organised and identified.

The opening and closing chapters follow a familiar form. Immediately after this introductory chapter there is a review of the awareness management and communication architectures of multi-user virtual reality systems (which are the form of CVEs being considered in this thesis). Chapter 2 also situates Collaborative Virtual Environments within the broader area of Computer Supported Cooperative Work (CSCW). This chapter provides background material, context and motivation for the main body of this work.

The closing chapter of this thesis comprises a summary of the work presented and its main contributions followed by suggestions for future work. The thesis concludes with more personal and informal reflections on the work described in this thesis and its relationship to computer science and society at large.

The central eight chapters are divided into two parallel parts. Each part deals with a single prototype system together with the theory which underlies it and the results which flow from it. The two parts represent two distinct phases of activity in the work presented here; each approaches the same issues but in different ways. This distinction is retained within this thesis to give a broader presentation of the problems considered and a clearer progression through the body of the thesis.

These eight chapters sit within a matrix of issues and ideas:

Key concept	Theory	Prototype overview	Social issues	Communication issues
Part I: Direct relationships	Chapter 3: Original spatial model	Chapter 4: MASSIVE-1 overview	Chapter 5: Awareness relationships	Chapter 6: Spatial Trading
Part II: Indirect relationships	Chapter 7: Third party objects	Chapter 8: MASSIVE-2 overview	Chapter 9: Regions and awareness	Chapter 10: Awareness driven communication

As shown, each part begins with a theoretical foundation. In the case of Part I this was a pre-existing theory, the so-called spatial model of interaction, which is reviewed in chapter 3 as necessary background material. The spatial model of interaction reasons about awareness between pairs of objects (and/or users) in direct relationships. In the case of Part II the theoretical foundation is the principle theoretical contribution of this work, namely the concept of third party objects, a powerful extension to the original spatial model of interaction which is presented in chapter 7. This extension builds directly on the original model, extending it to include the effects of context and environment on interaction, i.e. *indirect* relationships and effects. Chapters 3 and 7 may be read consecutively to gain a complete view of the spatial model of interaction and third party objects.

In each part of this thesis, the theory is followed by an overview of the corresponding prototype system; these are MASSIVE-1 and MASSIVE-2, respectively (where M.A.S.S.I.V.E. is a Model, Architecture and System for Spatial Interaction in Virtual Environments!). Both prototypes are multi-user distributed virtual reality systems, and implement most or all of the corresponding model of interaction. MASSIVE-1 is the earlier and more limited system. It is dedicated to virtual reality tele-conferencing which it supports through combinations of text, graphical and real-time audio interaction over unicast network protocols. MASSIVE-2 is the more recent and more general system. In addition to tele-conferencing, it provides a general application development framework and employs a combination of multicast and unicast network protocols. Chapters 4 and 8 introduce the prototype systems, describe the typical user features and provide an overview of the software system, its components and its organisation. The structure of the two systems is significantly different and so at least some familiarity with this material is helpful for the chapters which follow.

The remaining four chapters present the main design and evaluation components of this work. Each chapter has a similar structure, comprising:

- description of how the relevant theory has been implemented within the prototype;
- explanation and discussion of the characteristics and features which result;
- evaluation of the prototype with reflection on the underlying theory; and
- summary and conclusions.

These four chapters deal with the two major linked themes of this work (see figure 1.3 on page 12). Chapters 5 and 9 look "down" from the user's perspective within the everyday social world, and consider the kinds of interaction and control which are afforded to a participant in a CVE, i.e. the user's *desired* and *presented* awareness. Chapters 6 and 10 look "up" from the world of computers and networks, and consider the form and requirements of the communication which facilitates this interaction, i.e. the management and realisation of *requested* and *achieved* awareness in the terminology of figure 1.3.

Chapters 5 and 6 are based on MASSIVE-1 and the original spatial model of interaction. Chapter 5 considers the implementation and affordances of (spatial model) awareness. Chapter 6 considers the nature, utility and scalability of spatial trading, which is a formalisation of the "aura" component of the spatial model within a distributed systems framework. Chapters 9 and 10 are based on MASSIVE-2 and the third party object concept. Chapter 9 deals with the proposed extensions to the awareness model, while chapter 10 addresses distribution and in particular the use of multicast-based communication protocols in the context of the extended spatial model. As with chapters 3 and 7 there is a logical progression between the two parts, however each chapter can be read independently.

There follows a complete map of the structure of this thesis which the reader may wish to refer to as an aid to navigation within the text.

Chapter 1. Introduction

Chapter 2. CVEs: a review

Key concept	**Theory**	**Prototype overview**	**Social issues**	**Communication issues**
Part I: Direct relationships	Chapter 3: Original spatial model	Chapter 4: MASSIVE-1 overview	Chapter 5: Awareness relationships	Chapter 6: Spatial Trading
Part II: Indirect relationships	Chapter 7: Third party objects	Chapter 8: MASSIVE-2 overview	Chapter 9: Regions and awareness	Chapter 10: Awareness driven communication

Chapter 11. Summary, conclusions and future work

2. CVEs: a Review

This chapter presents a review of the field of collaborative virtual environments. This focuses on multi-user virtual reality and in particular two issues which are central to this thesis: scopes of interaction (which is equivalent to awareness management) and communication architecture. Section 2.1 situates this work within the broader context of Computer Supported Cooperative Work (CSCW). This situates CVEs (at least of the form considered here) in the area of "Same-time/Different place" collaboration, i.e. synchronous distributed interaction. Section 2.2 then reviews and explores the issues of interaction scoping and communication architecture in CVEs. The consideration of interaction scoping in section 2.2.1 identifies a number of dimensions within which current systems and approaches can be classified. It is apparent from this review that work is needed to develop support for social factors in interaction and expressiveness. These are key areas addressed in this thesis: support for social factors is central throughout this thesis, while expressiveness is a particular focus of part II. The review of communication architectures in section 2.2.2 provides necessary context and background for this thesis. In particular it considers the utility of the peer-to-peer networking approach which is adopted for both of the prototype systems described in this thesis and introduces and explores the use of network supported multicast communication which is a key element of part II.

2.1. CSCW

Computer Supported Cooperative Work is a relatively recent discipline within computer science; the term was first used in 1984 [Greif, 1988]. According to Schmidt and Bannon [1992, p.11]:

> CSCW should be conceived of as *an endeavour to understand the nature and requirements of cooperative work with the objective of designing computer-based technologies for cooperative work arrangements.*
> [their emphasis]

The technologies and products within this field are often referred to as *groupware* [Ellis et al., 1991]. A principle taxonomy which is applied to CSCW systems concerns the timing and location of the collaborative activity. This is shown in table 2.1, reproduced from [Ellis et al., 1991].

Table 2.1: dimensions of time and space applied to CSCW systems

	Same Time	**Different Times**
Same Place	face-to-face interaction	asynchronous interaction
Different Places	synchronous distributed interaction	asynchronous distributed interaction

CVEs, at least at the present time, focus on the "Same Time/Different Places" category, i.e. providing support for synchronous distributed collaboration. When emphasising the inter-personal communication involved this may also be referred to as tele-conferencing or real-time conferencing although there is a historical tendency to apply the former term exclusively to remote audio and video conferencing (e.g. [Sarin and Greif, 1985]). As well as CVEs there are other types of system which support synchronous distributed collaboration. These include multiuser editors, which emphasise shared document creation and modification (e.g. GROVE [Ellis et al., 1990] and Quilt [Leland et al., 1988]), and group decision support systems, which emphasise the process of collaboration around general problems (e.g. COORDINATOR [Flores et al., 1988]).

CVEs may be distinguished from these other types of CSCW systems by considering the relationship between a user and the shared aspects of the system. In CVEs the user is directly represented within the shared environment and their interaction with the shared environment is mediated by means of their embodiment which acts as a kind of real-time proxy for the user. Also, in CVEs the shared space in which the interaction occurs is malleable, i.e. open to reinterpretation and appropriation [Robinson, 1993]. This may be contrasted with a user's interaction with a synchronous shared editor, for example. Such a user may be embodied by their own cursor or tele-pointer which indicates (to some extent) what they are attending to within the shared document and where they are likely to make changes. However the shared space, in this case the document itself, has an *a priori* interpretation - letters, words, paragraphs, etc. Consequently, there is little scope for users to give new meanings to the space or to express the form of their interaction through it (e.g. through the kinds of "natural" spatial communication and peripheral awareness which were highlighted in section 1.1 as key motivations underlying the development of CVEs).

There are two main types of system which can legitimately claim to be CVEs: MUDs (Multi-User Dimensions/Dungeons) and multiuser virtual reality systems. These two types of systems have very different forms of presentation but have important similarities: users are represented and embodied within the virtual environment; users are independent; the environment has a (broadly) spatial structure (although in a MUD it is normally a graph rather than a Cartesian coordinate system); and the shared environment is a containing context rather than an end in itself. Although most MUDs are used for gaming and role-playing activities, there are examples of more "serious" use such as [Curtis and Nichols, 1994]. There are also interesting MUDs and MUD-like systems which include (typically 2D) graphics and navigation in a Cartesian coordinate system such as Habitat [Morningstar and Farmer, 1991].

This thesis - and the remainder of this chapter - focuses on multi-user virtual reality systems as the more significant and (in some sense) idealised realisation of a collaborative virtual environment.

2.2. Multi-user Virtual Reality

This section, which forms the remainder of this review of CVEs, focuses on two aspects of the realisation of existing multiuser virtual reality systems from the perspective of large-scale collaborative virtual environments: scopes of interaction and communication architecture. The focus of this review is primarily on issues of scalability in multiuser virtual reality. The issue of scoping interaction is considered first.

2.2.1. Scoping Interaction

It was observed in chapter 1 (section 1.2) that people have characteristic perceptual and cognitive capabilities and limitations with regard to perceiving and interacting with their physical environment. In a complementary way, many of the limitations to scale (e.g. in a user's local machine) need not depend on the total number of simultaneous participants. They may depend instead on the subset of current participants which are presented to a user, i.e. the scope of a participant's awareness of the virtual environment, or their scope of interaction [Macedonia et al., 1995]. This section reviews a number of existing multiuser VR systems and draws out a number of conceptual and implementational choices and issues in this area.

Firstly, a number of existing systems' approaches to scoping interaction are outlined below.

- AVIARY [Snowdon and West, 1994] is a strongly object oriented distributed VR system. Amongst other interesting features it employs a generic collision detection service to discover artefacts within (or close to) spatial regions of interest such as a visualiser's view frustum.
- WAVES [Kazman, 1993 and 1995] focuses on message passing with active message filtering services in a process and object based system. This includes message filtering based on viewpoint and area management (drawing on a simplified subset of [Funkhouser et al. 1992] - see also the RING system, below) and supported by specialised area management servers.

- NPSNET [Macedonia et al., 1995] tiles the world with hexagonal cells, each with its own multicast group. Each observer has their own Area of Interest (AoI) which identifies the cells which are potentially of interest to them, so observers need consider only near-by cells (and receive network traffic only from those cells). Cell size is motivated by consideration of the application domain (military exercise simulation).
- PARADISE [Singhal and Cheriton, 1996; Singhal, 1996] employs "projection aggregation" which combines a coarse adaptive cell-based scheme (based on a spatial octree) with further partitioning into logically or operationally related objects. This approach provides a high-level method for establishing (potential) interest based on general spatial area. This system also introduces abstractions - simplified overviews of numbers of objects - which may significantly reduce communication requirements at a (virtual) distance (these are also a feature of the work presented here).
- The Spline system [Barrus et al., 1996] composes a world from regions or "locales" which are loosely comparable to cells in the NPSNET system but which may have an arbitrary shape. Interaction and awareness is limited to the current locale and its immediate neighbours (which are explicitly named). Neighbouring locales may be linked by arbitrary geometric transformations.
- The approach adopted by Broll [1997] divides the world into a hierarchy of disjoint cells or zones of different sizes and shapes, each having its own communications infrastructure (e.g. multicast groups). When external to a cell the participant may see an (optional) external representation of the cells contents (c.f. abstractions).
- RING [Funkhouser, 1995 and 1996] scopes interaction and communication according to potential visibility in densely occluded environments (e.g. within buildings). This is based on a client-server system with an optionally distributed server performing message filtering and visibility calculation.
- A less spatial approach to world structuring has been incorporated into the most recent version of DIVE (version 3), based on the association of light-weight multicast groups with object hierarchies [Hagsand, 1996]. These allow the logical structure of the world to be used to scope communication. However the rules for exploiting and applying these facilities are left to individual applications at present. DIVE version 2 provided some limited (demonstration oriented) support for part of the spatial model of interaction.
- The Virtual Society project [Lea et al., 1997] has a client-server based architecture which employs auras (inspired by the spatial model of interaction) to scope interaction. Interestingly, aura is also used to control other aspects of data consistency between users as well as scoping awareness. A single server version is produced commercially; a distributed server is under development.
- NetEffect [Das et al. 1997] is also client-server based, but with a simpler scoping notion of closed "community" groups with limited concurrent access. This is supplemented in some (but not all) areas by additional scoping based on location in a spatial containment hierarchy related to world content.

- Division Ltd.'s dVS system [Division, 1996] (developed from [Grimsdale, 1991]) is based on a partially replicated distributed database. In version 3 it supports a top-level grouping notion of "zones" which may be enabled and disabled explicitly by an application to create hand-crafted forms of scoping (however zones have been removed from version 4).

Other multiuser virtual reality systems exist such as VEOS [Bricken and Coco, 1994], MR Toolkit [Wang et al., 1995] and BrickNet [Singh et al., 1994]. However they do not explicitly address issues of awareness scoping.

Each of the systems described above seeks to systematically limit the communication which occurs between the world and each participant (or machine). Consequently this limits the awareness which each participant has of the world as a whole. As has already been observed, this limitation of awareness and communication may be justified by the localisation of knowledge and interaction which occurs in the everyday physical world, where it is due to the physics of light, sound, forces, etc. in combination with typical perceptual and cognitive limitations of human beings. In each case the system may be viewed as providing a computational approximation to physical world constraints, either directly (as in NPSNET) or more metaphorically (as in DIVE).

A number of significant issues and choices emerge when considering the partitioning or restriction of awareness. The first, and perhaps most significant, is the basis for partitioning. Three main alternatives are listed below.

- Space - partitioning may be purely spatial, for example based on distance in "open" spaces (e.g. in NPSNET).
- Content - partitioning may be based on world content such as walls and buildings (e.g. in RING).
- Semantics - alternatively partitioning may be based on logical or organisational rather than spatial considerations (e.g. in DIVE).

As well as the basis for partitioning, there are a number of other significant issues or dimensions to be considered in the design or choice of an awareness partitioning scheme. These include: continuity of awareness; continuity of movement; support for heterogeneity; support for live modification; expressiveness; support for social factors; accuracy; and exploitability. These are described in turn below. Table 2.2 on page 23 and table 2.3 on page 24 indicate for each of the systems considered the degree of emphasis and support given to each of these areas.

- Continuity of awareness - can observers see into the distance, or are there arbitrary boundaries imposed on awareness? For example in Spline the observer cannot see beyond the next locale whereas in NPSNET the area of interest may extend to cover more distant cells.
- Continuity of movement - are there distinct jumps in awareness as new objects or regions come into range, and if so, at what granularity? For example there will be jumps associated with zone transitions in many of these systems.

Note that it is the degree and nature of continuity which distinguishes the partitioning schemes considered here from systems which divide awareness solely on the basis of multiple virtual worlds.

- Support for heterogeneity - to what extent are the partitioning choices made in common for all participants, and to what extent can individual participants tailor the information they receive (e.g. according to the power of their machine)? In these systems zones (where present) are fixed for everyone. However in NPSNET for example (and other systems based on aura or related notions) each participant may have their own variable-sized area of interest which may span varying numbers of cells or objects.
- Exploitability - can these nominal constraints on awareness be exploited to yield reduced bandwidth and state-handling requirements at individual machines? All of these approaches are exploitable (and exploited). However the efficiency of exploitation will depend on further factors such as accuracy (defined below).
- Support for live modification - can the awareness structure be modified within a running system with correct results, or is it fixed in advance? Where it is related to world content or semantics, does it change automatically in response to changes in these things, for example knocking down one building and erecting another?
- Expressiveness - can the system or mechanism express or represent a range of partitioning strategies? For example, can it combine distance constraints in open spaces (as in NPSNET) with bounded regions such as rooms and buildings (as in RING or Spline)?
- Support for social factors - do the systems provide awareness in a form which is sufficiently flexible and appropriate to support effects observed in real-world interactions? These include: peripheral awareness of "nearby" activities, focus of attention, balance of power, context sensitivity, spontaneous interaction and targeted production of information (these issues are covered in more detail in section 3.1). At present the MASSIVE systems as presented in this thesis are alone in being based on a sociologically motivated model of general awareness in interaction. This area is a key focus of this thesis.
- Accuracy - how accurately does the approach model the "ideal" awareness which participants might expect in a given situation? In general, using more cells or zones will allow greater accuracy, though probably at some cost in terms of management. This issue is considered at some length in chapter 10.

Of course there will be implementational costs in each of these areas measured in terms of factors such as efficiency, management overhead and complexity. Every system designer must make a trade-off between an abstract ideal and the realities of working systems. For example, NPSNET avoids live modification and has limited expressiveness, but achieves continuity of awareness, reasonable support for heterogeneity and reasonable accuracy within a particular application domain.

Some general areas of weakness can be observed in current systems, especially support for social factors and expressiveness (and, to a lesser extent, support for live

Table 2.2: issues of scoping interaction in current multiuser VR systems (i)

System	Basis	Continuity of awareness	Continuity of movement	Support for heterogeneity	Exploitable
AVIARY	space	good	good	good	good
WAVES	space or content	fair to good: might be extended	fair: large "areas"	fair: depth of traversal	fair to good
NPSNET	space	good	fair: large cells	fair to good: AOI	good
Paradise	space plus organisat'n	good	poor to fair: course octree	fair	good
Spline	space or content	fair: limited to next locale	fair: large locales	poor	good
[Broll, 1997]	space or content	fair: limited	fair: large cells	poor	good
RING	content	good	good	poor	good
DIVE	organisation	good	undefined	good but undefined	fair to good
Virtual Society	space	good	good	good	good
NetEffect	organisat'n and content	poor	poor	poor	fair
dVS	space or content	typically poor	fair: large zones	poor	fair: rendering optimization

modification in conjunction with multicasting and expressiveness). While these systems all have some form of awareness scoping, it is motivated by considerations of scalability and performance; the implicit use of space and awareness in communication is not considered. The work presented in this thesis emphasises support for social factors hand in hand with scalability. In addition, the later work (the third party object concept and the MASSIVE-2 system) emphasises issues of expressiveness and live

Table 2.3: issues of scoping interaction in current multiuser VR systems (ii) plus multicast support reproduced from table 2.4 on page 26

System	Live modify	Expressiveness	Social factors	Accuracy	Multicast
AVIARY	good	poor: open spaces	no	open spaces: good	no
WAVES	good	fair: linked spaces	no	linked spaces: fair	no
NPSNET	none	poor: open spaces	no	open spaces: fair	yes
Paradise	good	fair: open spaces plus organisation	no	open spaces: poor to fair	yes
Spline	fair: position, not shape	fair: linked spaces	no	linked spaces: fair	yes
[Broll, 1997]	none	fair: bounded spaces	no	bounded spaces: fair	yes
RING	none	poor: interior spaces	no	interior spaces: good	no
DIVE	good	undefined	v.3: no v.2: limited	undefined	yes
Virtual Society	good	fair: open spaces plus linked worlds	no	open spaces: fair	yes: servers only
NetEffect	none	fair: bounded spaces	no	bounded spaces: poor to fair	no
dVS	poor	fair: application-linked spaces	no	linked spaces: fair	no

modification. Chapters 6 and 10 in particular return to the theme of scoping awareness.

2.2.2. Communication Architecture

The previous section has set the agenda for this thesis: support for social factors and expressiveness in large-scale CVEs. This section provides supporting consideration of network and software architectures for this type of system. It reviews and reflects on the approaches to network communication adopted in the systems considered in the previous section. In particular it considers the use of peer-to-peer networking which is adopted for both of the prototype systems described in this thesis and introduces and explores the use of network supported multicast communication which is a key element of part II. This section provides background material for the prototype systems and evaluations presented in this thesis. The aspects of communication architecture which are considered here are:

- "patterns" of communication and distribution, in particular use of direct inter-peer communication and the role of centralised servers and services; and
- use of network multicast facilities.

In each case the main issues and choices are described and compared, with examples from the multiuser virtual reality systems referred to in the previous section. This is summarised in table 2.4 on page 26.

Patterns of Communication

The first issue considered is that of communication pattern, or the way in which communication is structured and organised within the system. A common distinction is between:

- peer communication, which takes place between comparable programs or processes with similar or identical roles; and
- client-server communication, in which one particular process (the server) provides a service to one or more other processes (the clients).

Distributed systems are likely to include elements of both peer-to-peer and client-server communication. For example, some elements of any distributed system naturally lend themselves to implementation and presentation as a service. Examples include name services and traders (such as the ODP/ANSA trader [ITU-T, 1995]). However there is more diversity in realising communication between comparable processes, such as between participants in a CVE. The main options are direct inter-peer communication or indirect communication via a common server. The attractions of direct peer communication are: typically reduced latency; the central server cannot become a bottleneck; there are no issues of single versus distributed server; and it is relatively straight-forward to make use of multicast communication facilities between peers. Conversely, the attractions of communication via a common

Table 2.4: communication architectures of multiuser VR systems

System	Peer communication?	Servers?	users in/out of system "core"?	multicast in core
AVIARY	yes	yes: system	in	no
WAVES	via message managers	yes: system	in	no
NPSNET	yes	no	in	yes
Paradise	yes	yes: system	in	yes
Spline	yes: in core	yes: http and system	in or out	yes
Broll, 1997]	yes: in core	yes: http and system	in or out	yes
RING	yes: in core (distributed server)	yes: system	out: servers in core	no
DIVE	yes	yes: http and system	in	yes
Virtual Society	yes: in core (distributed server)	yes: http and system	out: servers in core	yes
NetEffect	limited (one audio call)	yes: http and system	out: servers in core	no
dVS	yes	yes: system	in	no

server are: tailored and lower-overhead communication with each client; simpler communication for the client processes; and simple inclusion of other services.

Examples of CVEs which adopt client-server communication models are RING, Virtual Society and NetEffect. In the latter two cases this is directly motivated by the choice of a PC using a normal dialup modem connection as the principle delivery platform. The bandwidth limitations over the modem link are a critical consideration; this leads to a choice of a client-server architecture with tailored (e.g. compressed) communication between each client and the server(s). All three of these systems have a multi-server architecture in which a number of servers can cooperate to accommodate a larger number of clients. In the case of NetEffect this is a simple federation of servers, each catering for one or more complete worlds. In the case of RING the servers may act either as a fixed entry point to the core system, or may take responsibility for particular regions of the total virtual environment.

WAVES adopts a kind of hybrid architecture in which most communication is performed via general purpose message filtering processes which act as tailorable message routers.

The other systems all adopt a peer communication model, though with varying degrees of dependence on remaining server components. The Distributed Interactive Simulation [IEEE, 1993] approach has traditionally avoided any form of server, relying instead on extensive pre-configuration of systems. This is seen in the NPSNET and PARADISE systems. The remaining systems depend on some form of system-specific server, often in conjunction with standard HTTP servers. For example, AVIARY relies on a (potentially distributed) Environment Database (EDB) to maintain world state and to perform collision-driven introductions between objects.

The approaches adopted by each system are shown in table 2.4 on page 26. A good combined approach is to employ peer communication within the "core" of the system and client-server communication at the "edges". This is seen (in slightly different forms) in Spline, DIVE, the work of Broll, RING and in the distributed server of Virtual Society. In Spline, DIVE and the work of Broll the core system - where peer communication is employed - may include participants' machines. In the cases of RING and Virtual Society the core system is restricted to the components of the distributed server. However in every case participants with restricted network access depend on client-server communication to participate. In the cases of DIVE and the work of Broll this compensates only for non-availability of local multicast facilities rather than for limited access bandwidth.

The work presented in this thesis has not focused on delivery over low bandwidth or non-multicast access networks and so peer-to-peer communication has been the approach adopted in both prototypes. As in most of these systems additional server components are employed as appropriate.

Use of Multicasting

The second aspect of communication architecture which is considered in this chapter is the use of network multicasting facilities. Normal network communication is point-to-point, or "unicast". Each message which is sent is received by (at most) a single remote machine to which it was explicitly directed. In the case of multicast communication a single message may be sent to an abstract "multicast group" and received by any number of other machines or processes which are notionally members of that group. The actual copying of messages is handled by the network, typically through a combination of its physical characteristics (in the case of shared physical medium networks such as Ethernet) and by routers spread throughout the network. Figure 2.1 on page 28 illustrates the potential benefits of network supported multicasting when a sending machine (S) wishes to send the same message to four recipients. Note in particular the reduced number of packets on the WAN link and the shared LAN. Multicast communication has become available on UNIX operating system platforms in the past few years (and more recently on Microsoft's Windows 95 and NT platforms) in the form of IP multicasting [Deering, 1989; Deering and Cheriton, 1990]. This is supported over wide area networks by the experimental

MBone (Multicast back-bone) network [Macedonia and Brutzman, 1994]. Native multicast support is also becoming available on some commercial IP routers.

Fig.2.1: potential benefits of network multicasting

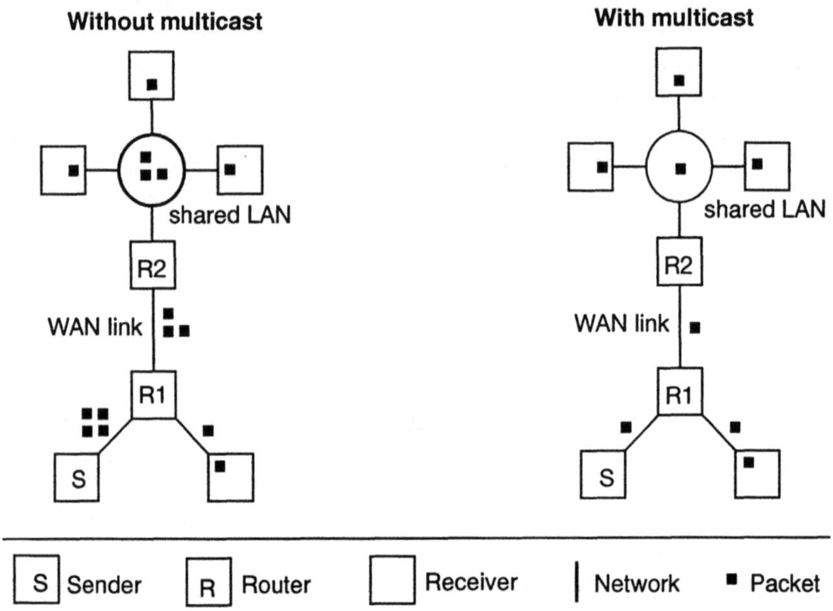

Multicast communication is extremely attractive for large-scale CVEs because it avoids carrying duplicate packets over the network. With unicast communication this would arise when, for example, one user was communicating with a number of other users over a common intermediate network (whether it be a shared LAN or an international link). With multicast communication only a single copy of the communicated data is sent over each common network and is copied as necessary to be distributed to the observers' machines. The primary advantage of multicasting is to reduce the total bandwidth requirements on shared networks and links. The disadvantages of multicasting are that: it requires support in the network which is not universally available at present; it requires some form of group management and group state in the network; and each group member receives the same messages (ignoring packet losses) even though they may have different detailed requirements. General networking issues such as achieving reliability, flow control and congestion detection and response are rather more difficult in the context of multicasting. It is also difficult to employ multicasting in a client-server situation because each client typically has its own unique (though possibly overlapping) requirements.

Of the systems which have been considered in this chapter the following employ multicasting: NPSNET, PARADISE, Spline, the work of Broll, DIVE and the Virtual Society distributed server. Where such a distinction can be made this is confined to the core of the system. "Second class" client-server based users in Spline, DIVE and the work of Broll rely on normal unicast communication (typically using the TCP/IP protocol). This is summarised in table 2.4 on page 26.

In systems which employ multicasting the multicast groups are typically associated with the same elements which scope interaction or awareness, primarily cells or locales. This reduces the number of multicast groups which are required compared to the total number of objects in the virtual world and so limits the multicast-related state and management overhead in the network and the end-machines. This approach is to be expected in current systems since scoping of awareness is largely a pragmatic concern of limiting network and machine resource requirements, rather than actively supporting and structuring communication and collaboration. One of the challenges addressed in this thesis is the combination of a sympathetic and appropriate use of multicast communication with more active and expressive management of awareness.

Of the two prototypes described in this thesis the first employs unicast communication while the second employs multicast communication. In terms of implementation this was one of the principle developments between the two systems. This can be seen in chapters 6 and 10. The potential use of multicasting was also an important consideration in developing and applying the third party object extension to the spatial model of interaction which is presented in chapter 7.

2.3. Summary

This chapter has reviewed the field of collaborative virtual environments, concentrating on multi-user virtual reality systems. These are (currently, at least) situated within the "Same time/Different place" domain of CSCW.

The focus of this thesis is on awareness as a facilitator of collaboration and scalability. This chapter has looked in particular at the ways in which interaction is scoped or managed in current systems; this performs the function of awareness management in these systems. It is apparent that support for social factors in interaction is a neglected area. Expressiveness is also an important area for work, as to a lesser extent is support for live modification.

Section 2.2.2 has considered issues of communication architecture. In particular, employment of multicasting is an important consideration for scalability in terms of total required bandwidth and also in terms of implications (constraints and facilities) for communication management. All of these issues and motivations are reflected in the work presented in this thesis.

The next chapter presents the spatial model of interaction as it existed at the outset of the work presented here and may be viewed as a continuation of this review.

3.
The Spatial Model of Interaction

This chapter describes the spatial model of interaction as it existed prior to the extensions proposed in this thesis. Section 3.1 describes some of the motivations behind the design and evolution of the spatial model of interaction. Section 3.2 presents the model itself. Section 3.3 describes other extensions to and developments of the spatial model which have been proposed prior to or contemporary with this work. Finally, section 3.4 illustrates the model with two previous demonstrations based on different components of the model.

This chapter is necessary background for chapter 4 which describes MASSIVE-1, a prototype multiuser virtual reality system which is based on the spatial model of interaction as presented in this chapter; this system is the first of the prototypes presented in this thesis and is the basis of the evaluations in chapters 5 and 6. This chapter also provides the context for the third party object concept proposed in chapter 7 and prototyped in MASSIVE-2 (chapters 8 through 10).

The principle statement of the spatial model of interaction can be found in [Benford and Fahlén, 1993]. This combines two strands, one led by Benford at Nottingham University [Benford et al., 1993] and one led by Fahlén at the Swedish Institute of Computer Science [Fahlén et al., 1993]. The main extensions (described in section 3.4) are due to Bowers [Bowers, 1993] and Rodden [Rodden, 1996]. Much of this work was developed within the COMIC project, a European Community ESPRIT III Basic Research Project which ran from 1992 until 1995.

3.1. Motivations

The motivations which lie behind the spatial model of interaction are essentially the same as the motivations for using and developing CVEs which were described in chapter 1, section 1.1. The spatial model of interaction is motivated by a number of specific considerations of work and space; these are listed below.

- The social and interactional significance of space. Space has a social significance which is important for real-world interaction: space can be viewed as a resource for managing activity and interaction ([Benford et al., 1995] which refers to [Giddens, 1984]). Also face to face communication involves body language, gesture, facial expression and gaze direction, all of which are situated within a spatial frame of reference.

- Supporting peripheral awareness. Studies of real-world activities in London Underground control rooms [Heath and Luff, 1991] and in air traffic control [Hughes et al., 1992] indicate that co-located workers maintain and make use of an ongoing awareness of the activities of others within their environment even when not explicitly cooperating. This is based on things like seeing "out of the corner of one's eye" or "at a glance" in conjunction with deliberately "making one's conduct available to others" [Bowers, 1993].
- Maintaining autonomy. In part, the spatial model of interaction reacts against attempts to formalise and control computer supported cooperative work which can be seen in some areas of CSCW such as workflow solutions (e.g. [Glance et al., 1996]). One of the spatial model's objectives is "minimising hard-wired constraints and replacing them with a model of increasing effort" [Benford and Fahlén, 1993]. Similarly it seeks to maintain parity or "balance of power" between speakers and listeners.
- Supporting flexible and dynamic group formation and development. In everyday group activity there is a tendency to group and re-group continually as the collaborative activity unfolds; the spatial model attempts to facilitate and represent this.
- Supporting informal and opportunistic interaction. Root [1988] and Gail [1991] observe that it is important to facilitate casual and social interaction in computer-mediated communication systems. This is analogous with, for example, meeting in the corridor or around the coffee machine in a more traditional working environment and is an important element of informal collaboration.
- Scaling to large numbers of participants. The spatial model of interaction includes concepts and facilities which allow effective management of interaction which, it is asserted, will become critical as the number of users increases [Benford and Fahlén, 1993]. This reflects considerations of both cognitive and computational overload. For example, the use of space provides a basis for structuring, exploring, mapping and navigating large virtual working environments.

Having described some of the motivations which have informed the spatial model of interaction the next section describes the model itself.

3.2. The Model

The spatial model of interaction assumes a space which is populated by potentially communicating objects. These objects may represent anything: human users, computer-based agents or data in a database for example. The space itself may have any form, for example a three dimensional Cartesian space, an abstract higher-dimensional space or a graph. The spatial model of interaction provides a framework for these objects to manage their interaction and communication and a key component of this management of interaction is their use of the space itself. Thus by controlling their position, orientation, distance, etc. the objects are able to modify their interaction and communication.

The model itself defines five linked concepts: *medium, awareness, aura, focus* and *nimbus*. These are extended by additional concepts of *adapters* and *boundaries*. The five basic concepts are dealt with in this section. Adapters and boundaries are described in section 3.3. First the concepts will be introduced and then their relationships and interactions will be considered.

- Medium. All interaction and communication occurs within a medium which defines what can be communicated. In a CVE a medium may be treated as a communication type. Typical media might include audio, video, graphics and text. There might also be more specialised object or application-specific media with particular roles or capabilities. A prerequisite for useful communication is that two objects have a compatible medium in which both objects can communicate.
- Awareness is perhaps the most important element of the model: it quantifies the degree, nature or quality of interaction between two objects. The spatial model may be viewed as a framework for negotiating values for awareness. These values then control the actual communication and interaction which takes place. Awareness is unidirectional and is specific to each medium, so two objects may have different levels of awareness of each other and different levels of awareness in different media. In principle every relationship between every pair of objects in every medium can be characterised by an awareness value, though in reality many of these relationships will have no concrete representation and so will effectively have zero awareness. In the spatial model, awareness is made possible by aura and is negotiated using focus and nimbus (defined below).
- Aura. Every object which wishes to communicate will have one or more auras, one in each relevant medium. An object's aura defines its overall region or scope of interest in a medium. The collision of two objects' auras is the fundamental enabler of interaction in the spatial model. As objects move through space their auras move with them; the environment tracks compatible auras (i.e. those with a common medium) and notifies the objects concerned when their auras collide. This notification allows those objects to establish direct channels of communication and to begin to negotiate awareness directly.
- Focus. This represents an *observing* object's interest in a particular medium. "The more an object is within your focus the more aware you are of it" [Benford and Fahlén, 1993]. This is one half of the basis of awareness; the other half is provided by nimbus (below). Focus may be expressed as a region or as a function defined over space or may be based on the attributes (spatial or otherwise) of the other object.
- Nimbus. This represents an *observed* object's projection in a particular medium. "The more an object is within your nimbus the more aware it is of you" [Benford and Fahlén, 1993]. This is the other half of the basis of awareness. Like focus,

nimbus may be expressed as a region or as a function over space or may be based on the attributes of the other object. Note that focus and nimbus are symmetric; this is the basis of balance of power between speakers and listeners in the spatial model. Nimbus is required in normal interactions, for example, to distinguish between interrupting or shouting, normal conversation, peripherally available information and privacy or security restrictions.

To recap, all interaction and communication in the spatial model occurs within the context (and constraints) of specific media. Aura, awareness, focus and nimbus are all medium-specific and a single object may have many of each (normally one per medium). Consider two objects, "A" and "B". When their auras collide this is noted by the environment. The two objects are informed about the collision and this information is sufficient to allow the objects to contact each other. They are then able to negotiate mutual awareness levels directly between themselves with no external intervention.

Aura is important within the spatial model because it scopes and hence limits the number of awareness relationships which must be considered. This is a key element of the potential scalability of a system based on the spatial model of interaction. Figure 3.1 on page 34 illustrates the operation of aura in enabling interaction according to medium and proximity.

Fig.3.1: auras enabling interaction

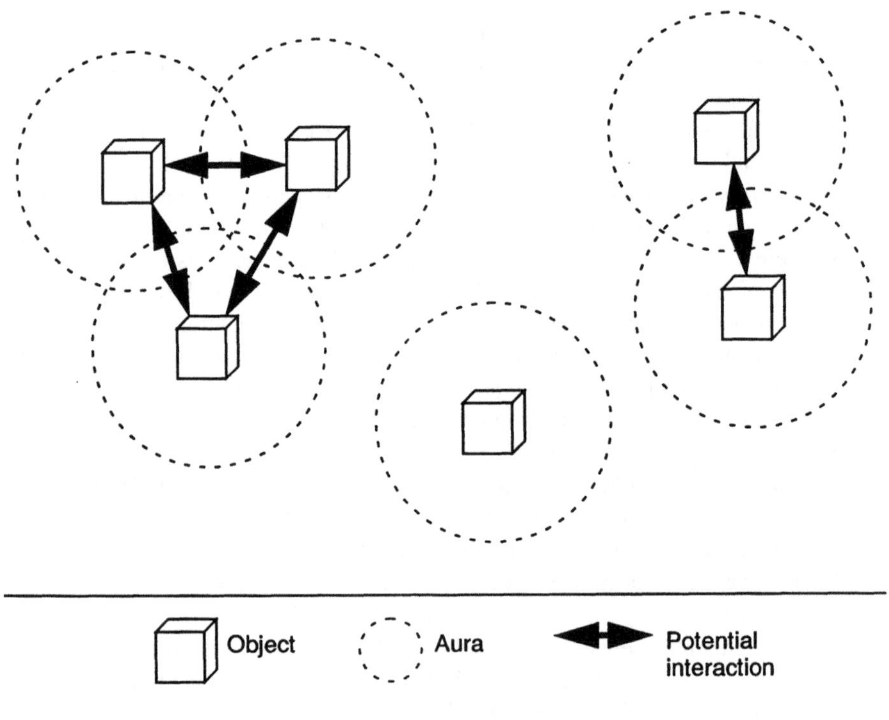

In the situation shown, two spatial groups have formed involving a total of 4 awareness relationships; without considering aura there would be 15 awareness relationships ($(n(n-1))/2$).

Once auras have collided the objects go on to negotiate awareness levels. Consider for example A's awareness of B. The negotiation process combines the observer's (A's) focus and the observed's (B's) nimbus. In the words of Benford and Fahlén [1993]:

> The level of awareness that object A has of object B in medium M is some function of A's focus on B in M and B's nimbus on A in M.

This awareness value which results is applied in a medium specific manner. For example, it might control the volume of an audio channel or the form of presentation (or otherwise) of a text message. Alternatively it might enable or disable access to particular facilities of services in the corresponding object. Figure 3.2 on page 35 illustrates the use of focus and nimbus in negotiating awareness values. This figure illustrates the potential of the spatial model to support autonomy, balance of power, dynamic group formation and scalability. Autonomy is provided because the objects are free to move and orient themselves independently within the space. Balance of power is due to the comparable influence of both focus and nimbus on the resulting awareness. Dynamic group formation follows naturally from individual mobility. Scalability is supported by the transmitter and receiver's ability to prioritise and allocate finite resources in an appropriate and (virtually) localised manner through choice of focus and nimbus.

Fig.3.2: two objects negotiating (unidirectional) awareness

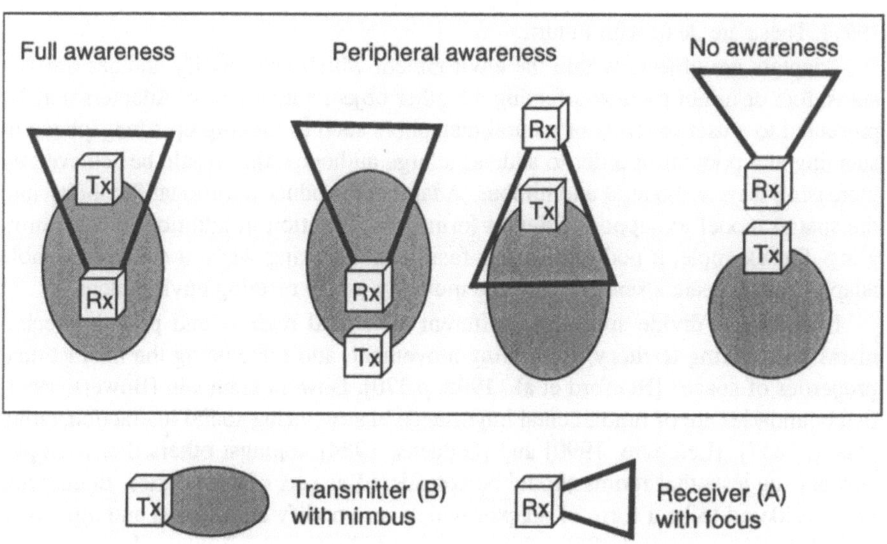

In the spatial model an object can control its awareness, and hence its interaction, by manipulating its own auras, foci and nimbi. This can be done in three ways [Benford and Fahlén, 1993]:

- implicitly, by moving and changing orientation within the space - its auras, foci and nimbi will follow the object as it moves about;
- explicitly, by directly modifying the parameters which define auras, foci and nimbi; and
- implicitly, through the use of adapters (which are described in the next section).

This section has described the core concepts of the spatial model of interaction. The next section describes extensions to and developments of the spatial model of interaction.

3.3. Extensions and Developments

The previous section described the core concepts of the spatial model of interaction: the way in which aura enables communication while negotiated awareness (based on focus and nimbus) controls and characterises that communication in an ongoing way. This section describes briefly two extensions to the spatial model and two models of awareness which are based on the spatial model. The extensions are *adapters*, from the original paper [Benford and Fahlén, 1993], and *boundaries*, from [Bowers, 1993]. The related awareness models are Rodden's application of the spatial model to shared object spaces [Rodden, 1996] and the Aether awareness model of [Sandor et al., 1997]. These are dealt with in turn.

Adapters are objects within the environment which can modify another object's auras, foci or nimbi thereby affecting the other object's interaction. Adapters may be presented to a user in terms of natural metaphors such as picking up a megaphone or standing at a podium in order to address a large audience; this would be achieved by increasing their audio aura and nimbus. Adaptors introduce additional flexibility into the spatial model to support different forms of interaction in addition to basic mingling. For example, a podium adapter facilitates lecturing while a conference table adapter could create a semi-private and mutually aware meeting environment.

Boundaries "divide space into different areas and regions and provide mechanisms for marking territory, controlling movement, and influencing the interactional properties of space" [Benford et al., 1995. p.370]. Bowers argues in [Bowers, 1993] that boundaries are of fundamental importance in structuring social interaction, citing [Marx, 1857], [Lefebvre, 1990] and [Giddens, 1984] amongst others. For example, Bowers suggests that rooms should be considered as sets of boundaries. Boundaries are considered to be a form of adapter which can modify aura, focus and nimbus in order to create a range of effects. In the context of boundaries Bowers also introduces the issue of control over movement which is not explicit elsewhere in the model. Unfortunately the mathematical details of realising boundaries are not considered.

Rodden [Rodden, 1996] presents a suggested generalisation of the spatial model which is defined in terms of shared objects rather than an explicit shared space. Awareness is calculated in terms of relationships with these shared objects (focus and nimbus). Conceptually the shared objects themselves constitute the space to which the model is applied. The theory is developed for linked (graph) spaces in particular, and is related to a number of CSCW applications: multiuser hypertext, workflow, versioning systems and shared desktops.

Aether [Sandor et al., 1997] is a related development of the spatial model which applies the spatial model concepts (principally focus, nimbus and awareness) to general semantic networks of objects and relationships. They propose that this would form a general low-level facility for creating CSCW systems. These semantic networks explicitly retain (suitably annotated) obsolete objects and relationships within the network, e.g. things which have been deleted or updated. Focus and nimbus are extended to include explicit specification of time (e.g. "the present", "the recent past", "two days ago"). The semantic network is subject to temporally managed garbage collection in order to limit its growth. The authors suggest that the inclusion of historical objects and relationships within the network breaks down the traditional division between support for synchronous and asynchronous collaboration. The other significant innovation in Aether is that the medium (which comprises a subset of the objects and relationships in the total network) actively modifies and transforms focus and nimbus as it is propagated through the network. This idea relates to the activities of third party objects proposed in this thesis (in chapter 7).

Having considered these extensions to the spatial model the next (and final) section of this chapter presents two partial demonstrations of the spatial model which have been reported in the associated literature.

3.4. Demonstrations and Examples

So far this chapter has described the spatial model of interaction, the motivations behind it and extensions which have been made to it. This final section describes two demonstrations of the components of the spatial model of interaction which have been reported in the associated literature. These are the aura-based audio and document facilities of the DIVE system [Fahlén et al., 1993] and an awareness-based text conferencing system [Benford et al., 1993].

The aura component of the spatial model originates from the work of Fahlén and others at the Swedish Institute of Computer Science (SICS). They describe how aura is used in their DIVE distributed virtual environment system to support audio interaction and document handling [Fahlén et al., 1993]. In this prototype aura is applied to a combined audio and document medium (but not to the graphical medium which is fully replicated within each virtual world). In this paper Fahlén et al. define aura as "the *nearfield* or immediate surroundings of a person"; effectively aura subsumes the functions of focus and nimbus.

The system includes two document tools: a whiteboard and a portable document. The whiteboard is comparable to a typical 2D graphical shared editor and supports the creation and manipulation of simple geometric objects by a number of simultaneous users. The whiteboard is a free-standing artefact within the shared virtual world. A document is a small portable single-user version of the whiteboard. A document can only be modified by its owner whereas the whiteboard can be used by anyone who comes within aura range of it. In terms of audio, when two users' embodiments come within aura range voice-talk is enabled and a communication channel is opened between them. The system also includes two adapter objects: a conference table and a podium. All users who are within aura range of the conference table are brought into a common conversational group. In effect their auras are adapted to match that of the table. A user standing on the podium (as determined by one of its two auras) is enabled to speak to a much larger group of users (which is determined by the podium's other aura). In effect the aura of the user on the podium is expanded to match that of the podium.

The focus, nimbus and awareness aspects of the spatial model were developed by Benford and others at Nottingham University. They describe CyCo (Cyberspace for Cooperation), a rooms based text conferencing system which might employ focus, nimbus and awareness to control interaction [Benford et al., 1993]. An additive model of awareness is proposed which is equivalent to the simple situation illustrated in figure 3.2 on page 35: when Tx is out of focus and Rx is out of nimbus then the awareness level will be 0; when Tx is in focus or Rx is in nimbus but not both then the awareness level will be 1; and when Tx is in focus and Rx is in nimbus then the awareness level will be 2. Focus and nimbus are simple discrete functions over space which have a value of 0 or 1 at any position; suggested shapes for focus and nimbus are (for a 2D environment) circles, sectors and rectangles (or infinite planes). This is implemented using the ANSAware Distributed Programming Platform [ANSA, 1989]. They also explain how continuous valued functions for focus and nimbus might be defined and employed, for example to control audio volume in an audio-capable system.

Both of these demonstrations are partial, predating as they do the integrated presentation of the spatial model in [Benford and Fahlén, 1993] and later work (including this). They are both limited in the media which they support: the first applies only to a combined document-and-audio medium, while the second applies only to text messages. They also lack any *explicit* user control over their interaction. However, taken together they do demonstrate the use of aura to enable interaction, the use of adapters to modify interaction, the distinct roles of focus and nimbus, and the use of a continuous valued representation of awareness (i.e. varying levels or degrees of awareness).

This completes the description of the spatial model of interaction as it stood prior to the work presented here. The next chapter, 4, describes MASSIVE-1, the first of the prototype CVEs which were created in the course of the work presented here. It describes the form and features of the system and its relationship to the spatial model of interaction as described in this chapter. This provides a basis and introduction for the particular evaluations presented in chapters 5 and 6.

4.
MASSIVE-1

This chapter introduces MASSIVE-1, the first of the CVE system prototypes presented in this thesis. This first prototype is based directly on the spatial model of interaction as described in chapter 3, i.e. as it existed prior to the work presented here. The goals of MASSIVE-1 were to:

- prototype and experiment with the spatial model of interaction;
- gain experience with CVE concepts and technologies; and
- create a usable virtual reality tele-conferencing system.

MASSIVE-1 was designed specifically to be a virtual reality tele-conferencing system. Some of its key features are:

- a full implementation of the spatial model which controls all interaction;
- rich computer mediated communication via combinations of text, graphics and packetised audio;
- a stable implementation running on low-end SG Indy workstations;
- the ability for text-only terminal users to participate in shared worlds; and
- support for multiple parallel worlds linked via portals.

It does not attempt to be a general-purpose virtual reality system. Typical features of a more general system which it lacks are: a well-defined API; non-specialist world authoring tools; direct manipulation of virtual objects; a systematic model of object behaviour; and support for popular graphical file formats.

This chapter provides a general introduction to MASSIVE-1 from the perspective of a normal user and also introduces the structure of the implementation. The most important aspects of the implementation are presented in more detail, with evaluation, in chapters 5 and 6. The first two sections of this chapter are primarily user-oriented. Section 4.1 describes the interfaces which are presented to a normal user (i.e. a participant in the virtual environment). Section 4.2 then describes the tools and facilities which are available within a virtual world. The last two sections deal with the implementation of MASSIVE-1. Section 4.3 describes the distributed programming model adopted. Finally, section 4.4 gives an overview of the software and network architecture of the system.

4.1. User Interface

This section describes the way in which MASSIVE-1 presents itself to a normal user. This includes the appearance of the system and also the types of control and interaction which are made available to the user.

MASSIVE-1 supports communication and interaction between users via a combination of 3D graphics, real-time audio and text. Each of these three forms of interaction is realised as a spatial model medium. Each medium is handled by a specialised user client process which provides a medium-specific interface between a user and the virtual world. A user can employ almost any combination of the three client programs; the only restriction is that they cannot use the audio client on its own, since it has no built-in support for specifying movement within the virtual world. So one user may be using all three client programs on a graphical workstation to give real-time graphical and audio interaction supplemented by text-based mapping and messaging. Another user may be logged in over the network from a VT100 alphanumeric terminal and have access to the text medium alone. These two users will be able to interact through the common text medium. Tools within the world may provide additional support for cross-medium interaction, for example, the text-to-speech convertor of section 4.2.

A user designates one of their client programs to be the "master", and this controls any other client programs which they may be using at the same time (the "slaves"). The master coordinates the activities of all of the user's client programs so that they present a consistent view of the virtual environment. The three client programs will be described in turn, starting with the graphical client, followed by the audio client and finally the text client.

4.1.1. Graphical Client

The graphical medium client maintains a single 3D graphical view of the virtual world; an example is shown in figure 4.1 on page 43. Each user is represented within the graphical medium by a simple embodiment (a "blocky") which is sufficient to convey the user's position and orientation and an indication of their identity (by means of a name label and customised body colour). In addition, the blocky indicates which media a user has access to. For example, a blocky with "ears" is audio-capable, a blocky with one "eye" is a desktop (monoscopic) graphical user, while a blocky with a "T" on their forehead is a text-only user.

The graphical medium client can be a user's master client or a slave client. When acting as a slave (to either a text client or to another graphical client) it just provides a view of the virtual world. However, when the graphical client is the master it pro-

Fig.4.1: graphical medium client display

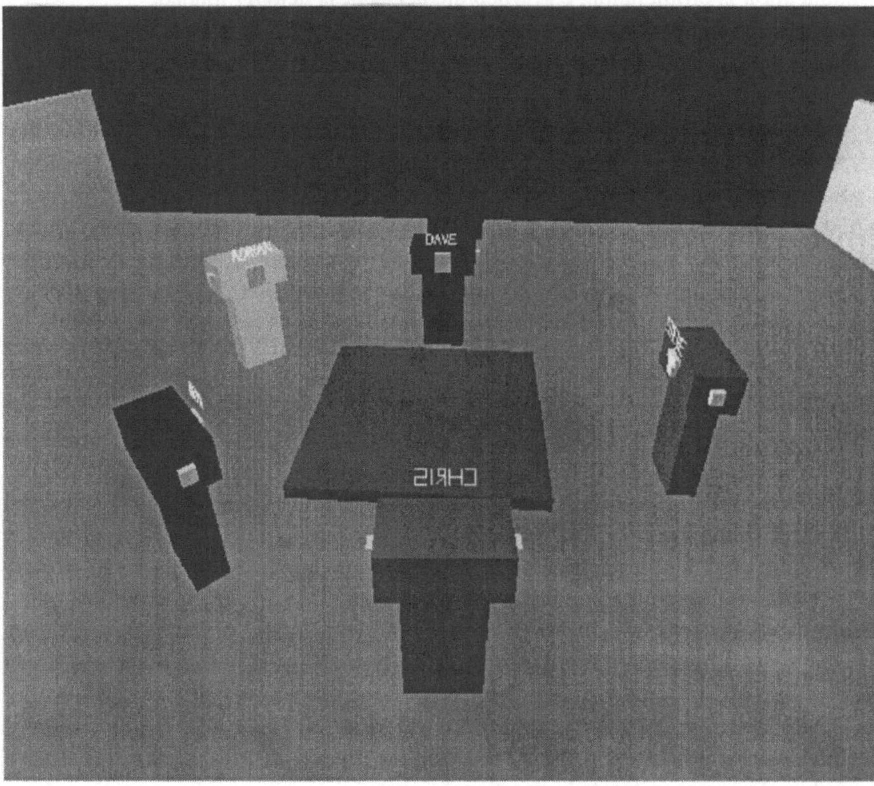

vides the user with a number of navigation and interaction control facilities which are listed below.

- Variable speed movement in six degrees of freedom. This is controlled using the mouse in different parts of the screen with combinations of mouse buttons.
- A choice between three settings for focus, nimbus and aura. These are "wide", "normal" and "narrow" and provide broad undirected interaction, mid-range semi-directed interaction and close-range highly directed interaction, respectively. These are stepped through using a single key press.
- The ability to continuously vary the angle and range of focus and nimbus. Like normal navigation this makes use of the mouse, but in combination with control keys. This is a relatively specialist facility.
- A choice between a number of simple graphical gestures. These include arm movements, pointing and "sleeping" (used to indicate that a user is not attending to the virtual world at present). These are selected using single key presses.

- A moving "mouth". This appears on the embodiment when the user is speaking, as a visual cue to speaker identity. This also acts as a diagnostic aid if audio communication is problematic, e.g. when the network is heavily loaded.
- An optional indication of the user's focus and nimbus. This is represented by a wireframe cuboid which approximates the region of maximum focus and nimbus.

Additionally, whether the graphical client is the master or a slave, it allows the user to choose (using key presses) between a number of pre-set viewpoints specified relative to their embodiment. The normal choice includes: the view out of their embodiment's eyes (the default); a view from above and behind their embodiment which shows other nearby objects; a view from overhead looking down on their embodiment which is effective as a map; and a view from in front looking back at their own embodiment. For each view the user can use keys or the mouse to zoom in and out.

The graphical client is the normal master client, but requires a reasonable performance graphical workstation such as an SG Indy.

Portals

One of the background concepts of the spatial model is that of a space or "world" within which objects and communication are situated. MASSIVE-1, like some other multi-user VR systems (e.g. DIVE [Hagsand, 1996]) includes "portals": a portal is an object in a world which forms a link or gateway to another world. As users move about within a world they can step "into" a portal and be transported to a new world and location, or to a different location within the same world. A portal's destination is specified when it is created by the world designer. Portals are unidirectional (but may be combined in pairs to create bidirectional links).

4.1.2. Audio Client

The audio medium client exchanges awareness and configuration information in the audio medium and uses this information to establish real-time audio connections between pairs of users and between users and other audio-capable objects. Audio in MASSIVE-1 is single channel u-law PCM (Pulse Code Modulation) encoded data at 8KHz; this is also referred to as "toll-quality audio" and is approximately the same quality as a domestic telephone call (but with significantly longer end to end delay). The audio client establishes audio connections only when awareness exceeds a threshold level. It also controls the playback volume to reflect the level of awareness so that sources heard with low awareness values are quiet while sources heard with high awareness values are loud.

The audio client manages a separate audio server process for each user. This was created specifically for MASSIVE-1 because existing network audio tools (such as VAT [Jacobson, 1992] and RAT [Hardman et al., 1995]) did not allow sufficient external control, for example of per-source playback volume. The audio client always operates as a slave client under the control of a text or graphical client. This is because navigation and other aspects of system control cannot be achieved via the audio client (which has no speech recognition facility).

4.1.3. Text Medium Client

The text client provides a simple map view of the surrounding area and allows the user to send and receive simple text messages. Figure 4.2 on page 46 shows a screen shot of the text client during a meeting; this is the same scene as in figure 4.1 on page 43.

The display has four components.

- The status bar at the top shows the orientation, location and focus/nimbus mode of the user.
- The column down the right of the screen identifies the objects in aura range and shows mutual awareness values.
- The character-based map in the centre shows the user's immediate surroundings in the virtual world. Objects are represented by letters with the key in the column down the right of the screen (the user's own embodiment is shown to themselves as an "@" symbol). User orientations are indicated by dashes adjacent to the appropriate character.
- The text window at the bottom of the screen displays recent text messages and allows the user to compose their own messages.

The text client can be a user's master client or a slave client. When it is the master client it allows the user to move about using key presses, change between settings for focus, nimbus and aura and perform simple "gestures" (which in the text medium are short preset text messages). The text client always allows the user to type text messages which are distributed to other users. Distribution and presentation of text messages depends on awareness level as illustrated in table 4.1 on page 46. At very low levels of awareness nothing is observed (the message is not seen). At intermediate levels of awareness an observer sees that something is said, but does not see its contents. At higher levels of awareness an observer sees the full message but it is shown in brackets to indicate that the message is not part of a focused (and nimbused!) exchange. At the highest levels of awareness the full message is displayed.

In terms of user machine capabilities the text medium client is the "lowest common denominator" and can be used on a text-only terminal such as a VT100. A text-only user can still be represented within the graphical medium but they cannot perceive anything in that medium (their graphical embodiment has a "T" on its forehead to indicate this to other graphical users). The text medium is useful in combination with the audio and/or graphical media as it provides a map view of the surroundings, and supports reliable messaging which is resilient to network congestion (unlike real-time audio).

Fig.4.2: text medium client display

```
Face: se  At:  -2.0,  1.0,  2.5     Focus: table
+------------------------------------------------------------+
|                                                            |
|  c   c   c   c   c   c   d   d   d   c   c   c   c   c    |c: conference room
|                                                            |
|  c   .   +   .   .   .   d   *   d   .   .   .   +   .    |t: table
|                                                            |
|  c   .   +   .   .   .   d   d   d   .   .   .   +   .    |d: door
|                                                            |
|  c   *   *   *   *   *   *   *   *   *   *   *   *   *    |C: Chris
|                                                            |  O-> 1.0, O<- 1.0
|  c   .   +   .   .   .   .   *   .   .   .   .   +   .    |S: Steve
|                                                            |  O-> 1.0, O<- 1.0
|  c   .   +   .   .   .   .   D   .   .   .   .   +   .    |R: Rob
|                      @         |                           |  O-> 1.0, O<- 1.0
|  c   .   +   .   .   . \ t   t   t   .   .   .   +   .    |D: Dave
|                                                            |  O-> 1.0, O<- 1.0
|  c   .   +   .   .   .   t   *   t - S   .   .   +   .    |
|                          /                                 |
|  c   +   +   +   +   + R t   t   t   +   +   +   +   +    |
|                                |                           |
|  c   .   +   .   .   .   .   C   .   .   .   .   +   .    |
|                                                            |
|  c   .   +   .   .   .   .   *   .   .   .   .   +   .    |
|                                                            |
|  c   .   +   .   .   .   .   *   .   .   .   .   +   .    |
+------------------------------------------------------------+


Chris says we'll be using audio for most of the meeting today
Rob says fine

Say no problem█
```

Table 4.1: effects of awareness level on presentation of the text message "hello" from user "Chris"

Awareness of Chris	Text Displayed
less than 0.4	nothing
0.4-0.6	"Chris says something"
0.6-0.8	"(Chris says hello)"
0.8-1.0	"Chris says hello"

4.2. Tools

The previous section described the facilities which users of MASSIVE-1 have to interact and communicate within a shared virtual world through the graphical, audio and text media. To supplement this direct communication the system also includes a number of active objects which may be placed within a virtual world to create opportunities for other types of interaction. These active objects are a message board, a text to speech convertor and spatial model adapters. These tools are described in turn.

4.2.1. Message Board

The message board is a specialised process which interfaces with both the text and graphical media. It "listens" in the text medium for message which are directed to it. It determines which messages are directed to it by checking the awareness level which the message board has of the message's sender. This takes into account both the message board's focus, which picks out senders that are in front of the board and relatively close to it, and the sender's nimbus, whatever that may be. The message board displays in the graphical medium the last few messages which it has received at high awareness. So users can add text to the board by approaching it (to increase its awareness of them) and typing a message in their text client. The board makes these messages available in a more persistent form in the graphical medium. A meeting around a message board is shown in figure 4.3 on page 48.

4.2.2. Text to Speech Convertor

The second tool, the text to speech convertor, has many similarities to the message board, above. It monitors the text medium for messages which are directed to it and reproduces them in the *audio* medium. The text to speech convertor makes use of a freely available (low quality) software speech synthesizer, interfacing it to the text medium for input and repackaging and re-sending its output in the audio medium. The text to speech convertor gives feedback to a user by changing its appearance when it is aware of them (a simple representation of a face is displayed). This appearance is specific to each observer and is an example of "subjectivity" in a virtual environment (see [Snowdon et al., 1995]).

The message board and the text to speech convertor both demonstrate how the spatial model can be used by agents wishing to take an appropriate role in multi-user, multi-agent settings. Using negotiated awareness levels these tools to are able to distinguish between background conversation and directed statements which might be commands or requests. The tool's awareness of a text message depends on its own focus, i.e. where the sending participant stands in relation to the tool, and the participant's nimbus, i.e. how they are projecting themselves in the space. Thus a tool can be made to respond only to text messages which are directed to it (participant's nimbus) and which come from participants standing just in front of the tool (tool's focus). This theme is explored further in [Benford and Greenhalgh, 1995].

Fig.4.3: the message board in use

4.2.3. Adapters

MASSIVE-1 supports adapter objects as proposed in [Benford and Fahlén, 1993]. Adapters are objects which change (or replace) a user's auras, foci and nimbi in order to transform the way in which they interact and communicate. The two adapters which have been used in MASSIVE-1 are a podium and a conference table (c.f. [Fahlén et al., 1993]). When a person stands on the podium they are given large auras and nimbi so that they can be seen and heard at a greater distance; this facilitates lecturing and similar patterns of communication. When a person approaches the conference table their auras, foci and nimbi are transformed to encompass the table and its immediate surroundings, but to restrict interaction further away; this creates a self-contained mutually aware group when participants gather around the conference table.

Adapters in MASSIVE-1 - like those in the spatial model of interaction - affect individual objects, changing their auras, foci and nimbi (this may be contrasted with the form of adaptation found with third party objects in chapter 7, which acts directly on the *relationships* between pairs of objects). Adapters are triggered by proximity which is determined by aura collision in a specialised "adapter medium". It is simple to create new adapters in MASSIVE-1 which have different parameters for aura, focus and nimbus. However it is not possible to combine adapters or to move or carry them.

4.2.4. Summary

This section has described a number of tools which enrich communication within virtual worlds. The message board enhances interaction between text and graphical users and introduces an element of support for asynchronous interaction (e.g. leaving messages for others to find at a later time). The text to speech adapter further enhances the potential involvement of text users in an otherwise audio-visual world; the addition of speech to text conversion would "complete the loop". Adapter objects allow different forms of interaction such as lecturing or focused group discussion in particular spaces. This concludes the description of the external characteristics of MASSIVE-1. The next two sections describe the implementation of the system, specifically the distributed programming model adopted and the overall communication architecture.

4.3. Distribution Model

This section describes the model of distributed computation which was adopted for MASSIVE-1. The next and final section of this chapter builds on this basic framework to describe the system's overall communication architecture.

CVEs are by nature distributed systems. There are many alternative approaches to distribution, see for example [Andrews, 1991]. Approaches which have been adopted for multi-user virtual reality systems include distributed databases (e.g. dVS [Grimsdale, 1991]), shared blackboards (e.g. VEOS [Bricken and Coco, 1994]), message-oriented models (e.g. DIS [IEEE, 1993]) and distributed object systems (e.g. WAVES [Kazman, 1993] and AVIARY [Snowdon and West, 1994]).

The style of distribution used in MASSIVE-1 is closest to a connected component model. In this approach the units of distribution are components - comparable to distributed objects - and communication is expressed in terms of message-carrying *connections* between those components (analogous to an electronic circuit with components and wires - see for example Regis/Darwin [Magee et al., 1994]). However, (like Microsoft's Common Object Model [Brown and Kindel, 1996]) the components in MASSIVE-1 are not explicitly defined but are represented indirectly by one or more typed interfaces (c.f. ODP [ITU-T, 1995] and CORBA [Vinoski, 1997]). The unit of distribution is a heavyweight operating system process and a single process can host any number of interfaces. Interfaces must be connected together before communication becomes possible. Figure 4.4 on page 50 shows two processes which are linked by a pair of connected interfaces.

Fig.4.4: two processes with connected interfaces

Depending on the types of the two interfaces (which need not be the same) each process can make remote procedure call (RPC) requests, send asynchronous notifications of events and share state in the form of attributes. A single local interface may be connected to many remote interfaces provided that the remote interfaces are all of the same type. Actions must be directed to a single connection but attributes may be associated with either a single connection (giving tailored presentation to each remote process) or with the interface as a whole (being shared by all connections to that interface).

The distributed programming model used (and implemented) here diverges from a more message-oriented distributed object model (such as ODP) because of shortcomings in that approach, namely:

- the difficulty and awkwardness of simulating features which are relatively easy with connections such as having a well-defined context for ongoing interaction, and achieving fate-sharing between peer-specific aspects of two associated objects; and
- the poverty of the attribute model (simply corresponding to a get and a set operation) compared with, for example, a distributed database which would include locking and the option of asynchronous notification of changes.

The distribution system and communication libraries were written specifically for MASSIVE-1, however the design could also have been realised in a number of commercial or research-oriented distributed systems though with varying degrees of awkwardness. The exact implementation details have little impact on the work presented in this and the following two chapters. The paper [Greenhalgh, 1994] includes some additional implementation details of MASSIVE-1 and the underlying system while [Greenhalgh, 1996] includes some more general reflections on distribution paradigms and the evolution of MASSIVE-1 and MASSIIVE-2.

4.4. Implementation Overview

This final section introduces the overall network software architecture of MASSIVE-1 using the communication model described in the previous section. More details of the most important aspects of the implementation are presented, with evaluation, in chapters 5 and 6.

Summarising from section 3.2, the background against which the spatial model is defined comprises worlds - which are disjoint spaces within which communication can occur - and objects - which are present within worlds and which are potential producers and consumers of information (which may represent human participants or computer programs). The key concepts of the spatial model itself are medium, aura, awareness, focus and nimbus (supplemented by adapters and other tools). In the design of MASSIVE-1 these various concepts are mapped onto and realised in terms of two fundamental relationships:

- the relationship between an object and the world in which it is present; and
- the relationship between two objects.

The former relationship deals principally with the medium and aura components of the spatial model of interaction, and is explored in detail in chapter 6 from a computational and networking perspective. The latter relationship deals with awareness, focus and nimbus and is dealt with in chapter 5 from a more social and user-based perspective. The remainder of this chapter introduces these two relationships and shows how they fit into the total system.

Figure 4.5 on page 52 shows an overview of MASSIVE-1's communication architecture, focusing on a single user, *A*. Each box is a separate process and each interface is labelled with its type (the unlabelled interfaces have a null type).

On the left are the processes which are local to user *A*: their master client (which must be a text or graphical client); a representative slave client (e.g. an audio client); and a trader process which is a simple attribute-based trading service as found in ODP. Slave clients use the trader to locate the appropriate master client based on the controlling user's name. Each client creates a local *traderc* (trader client) interface and connects this to the trader's *traders* (trader server) interface which has a well-known, preconfigured address. The master client registers interface offers with the trader which are subsequently passed on the slave clients in response to their requests. Each master client has a *location* interface and a *presence* interface to which its slave clients connect (via null interfaces). The *location* interface communicates the user's current world, location and orientation to the slave clients while the *presence* interface conveys information about the user's current aura, focus and nimbus settings.

On the right are the remote processes: an aura manager process and a representative remote client (e.g. another user's client process for the same medium). Every client process has its own *aura* interface with which it connects to an aura manager process (for simplicity *A*'s slave client's *aura* interface is not shown in figure 4.5 on

Fig.4.5: overview of MASSIVE-1's communication architecture focusing on user *A*

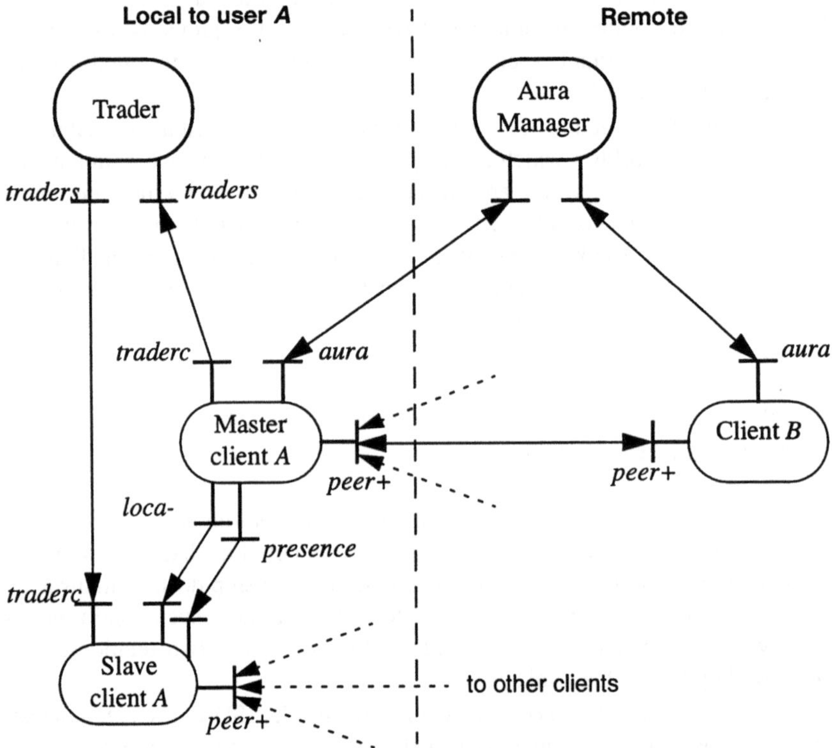

page 52). Many aura manager processes may exist, each being responsible for a different set of virtual worlds - the process of locating the appropriate aura manager is described in chapter 6. The aura interface includes attributes which describe the client's spatial model aura to the aura manager. This information includes the name of the world, the medium and the location and size of the aura. The aura manager continuously checks for collisions between auras in the same world and medium, and notifies the processes concerned via their *aura* interfaces.

When two process have been notified of an aura collision they establish a direct connection between the corresponding *peer* interfaces (the locations of these interface are passed on by the aura manager when the collision occurs). The *peer* interface includes the core facilities for negotiating awareness. The actual interface used (denoted by *peer+* in figure 4.5 on page 52) will be a subtype of *peer* which includes additional medium-specific operations and attributes (e.g. geometry for the visual medium). The peer processes use this direct connection to negotiate awareness levels and to exchange medium-specific information. As illustrated in the figure the same *peer* interface may be connected to many other processes as a result of multiple concurrent aura collisions.

To recap, when a user joins a MASSIVE-1 session they begin with a master client. This registers with the local trader and then contacts a well-known aura manager process and passes on (and keeps up to date) its world, medium and aura information. Additional slave clients may also be started, which use the trader to find and attach to the master client, and then contact the aura manager in the same way. All objects (such as user client processes) which are in the same world will be connected to the same aura manager; this performs aura collision tests and notifies them of aura collisions as they occur. The objects then establish direct peer connections which they use to negotiate awareness levels and to interact in specific media. In the case of user clients this information is presented to the user as described in section 4.1. When objects move out of aura range the aura manager notifies them of this and they tear down the direct connection, removing the other object from their local view of the environment (in that medium).

The final thing to note about MASSIVE-1 concerns provision of non-user world content such as rooms, gateways, adapters and other tools. Although much of the design of the spatial model and MASSIVE-1 is motivated by consideration of communication between people it is not restricted to this. The model is framed in terms of "objects" which may be user embodiments in the virtual world or could as easily be background scenery, software agents or aspects of an application's user interface. Aura, awareness, focus and nimbus are still relevant concepts. For example, a chair's focus might be zero but it still has a nimbus in one or more media which allows other objects (such as users) to be aware of it.

Non-user objects join and interact with the aura manager and with remote peers in exactly the same way as user clients, described above (except that they do not need to make use of the trader). They use the same *aura* and *peer* interfaces and maintain the same aura and awareness relationships. To reduce the number of processes required by the system (and the corresponding use of system resources) MASSIVE-1 has a standard world server process which reads a configuration file and creates and maintains appropriate *aura* and *peer* interfaces for a number of objects in the virtual world. The objects created by this process are mainly passive although they all support awareness negotiation. Objects can have representations in all three main media (text, graphics and audio) or a subset of them. Objects can also function as portals or adapters. The only medium-specific behaviours supported by the world server are realtime audio (generated from audio sample files) and selection between a number of graphical appearances, both according to the observer's awareness of the object. The message board and text to speech convertor have more complicated behaviours and are implemented as independent processes. However the method of interaction is the same.

This concludes chapter 4 which has introduced MASSIVE-1. The next two chapters examine and evaluate key components of its implementation of the spatial model of interaction. Chapter 5 considers the awareness relationship and the way in which this is presented to the user. Chapter 6 considers the aura relationship, which is formalised as "spatial trading", and the corresponding network resource requirements. Given the broad scope of this thesis some details of the implementation have been omitted for the sake of brevity. The interested reader can find additional details in [Greenhalgh, 1994].

5.
Direct Awareness

Chapter 4 introduced the MASSIVE-1 prototype's functionality and overall communication architecture. In particular, section 4.4 described how the spatial model of interaction is realised in terms of two key relationships: the awareness relationship, which links two interacting objects, and the aura relationship, which links an object to a world. The awareness relationship is the context within which objects (and participants) negotiate direct awareness based on focus and nimbus. This is where the facilities of the spatial model to support and manage interaction are realised and this is the area which has the most direct impact on users. This chapter assumes the existence and operation of the aura relationship (defining worlds and placing an outer limit on interaction), however the details of the aura relationship and the corresponding implications for network communications are deferred until chapter 6.

Section 5.1 describes how the awareness relationship is implemented and how direct awareness levels are calculated in MASSIVE-1. Section 5.2 then describes how awareness is used and controlled in relation to a user. Section 5.3 evaluates the effectiveness and limitations of awareness as realised in MASSIVE-1. Finally, section 5.4 summarises this evaluation and draws out a number of conclusions relating to awareness in CVEs, CSCW and the spatial model of interaction.

5.1. Implementation

When two compatible auras collide each of the *aura* interfaces involved (and hence the objects which they represent) is notified of the identity of a designated *peer* interface of the other object. One of these objects, chosen by the aura manager, establishes a new connection between these two *peer* interfaces, allowing the objects to communicate directly. This connection embodies and maintains the awareness relationship between those two objects. The relevant portion of figure 4.5 on page 52 is reproduced in figure 5.1 on page 56.

Section 5.1.1 describes the common *peer* interface type and its support for negotiating awareness. Section 5.1.2 describes how awareness is actually evaluated, including the form and parameters of focus and nimbus. Section 5.1.3 then explains how adapters are implemented.

Fig.5.1: an awareness relationship formed between two clients

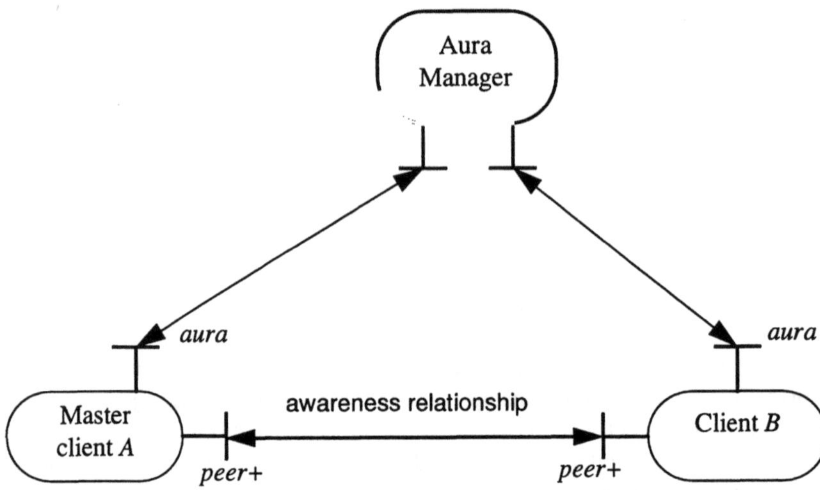

5.1.1. Peer Interface

The direct peer connection between two objects is medium specific and has a medium dependent type. However the spatial model is calculated in the same way in every medium. Every medium-specific interface type has a common supertype, the *peer* interface type. The awareness relationship is symmetrical and so the connection has the same type of interface on each end. The elements of the *peer* interface type are listed in table 5.1 on page 57. Note that some of the attributes are "objective" and apply to all awareness relationships which involve this interface (i.e. for that object in that medium), whereas some, specifically awareness, focus and nimbus, are "subjective" and apply individually to each awareness relationship.

The *identity* attribute uniquely identifies the object concerned (rather than the interface) and allows peer connections to be matched to collision events and allows the resolution of potential conflicts between multiple peer connections. The *position* attribute conveys the current location of the object within the (shared) world and is a key input for calculating focus, nimbus and hence awareness. *Radius* is the approximate spatial extent of the object and is particularly important when evaluating highly directional forms of focus and nimbus such as visual field of view. For example, approximating a room as a dimensionless point is rather misleading when you are standing inside the room. The remaining attributes are specific to a single awareness relationship and convey the object's current awareness of and focus and nimbus on the other object.

Table 5.1: the *peer* interface type

Name	Kind	Type	Objective/ subjective
identity	attribute	identity_t	objective
position	attribute	float[3]	objective
radius	attribute	float	objective
awareness	attribute	float	subjective
focus	attribute	float	subjective
nimbus	attribute	float	subjective

Consider a pair of objects, A and B, connected via *peer* interfaces as in figure 5.1 on page 56. The awareness negotiation to determine A's awareness of B is realised as follows.

- A takes B's position and radius from the connected interface and feeds them, together with its own position and orientation, into its focus function to yield a single floating point value which is its current focus on B. This value is used to update (if necessary) A's focus attribute for the connection to B.
- B does the converse, providing A with the current value of B's nimbus on A.
- A combines the value of its focus on B with the value of B's nimbus on A to calculate its overall awareness of B. This value updates (if necessary) A's awareness attribute for the connection to B.

B's awareness of A is negotiated simultaneously over the same connection. Within this framework each object can employ its own focus and nimbus functions corresponding to different forms and sizes of focus and nimbus. It can also use its own function for combining focus and nimbus to give awareness. The three constraints imposed by this particular realisation are: focus, nimbus and awareness must be single values rather than vectors; focus and nimbus must be separately calculable; and focus and nimbus can only depend on the remote object's location and radius. These constraints reflect a view of focus and nimbus as scalar fields over space; this is described in the next section which explains how focus, nimbus and awareness are evaluated.

5.1.2. Evaluating Awareness

The previous section described how values for focus, nimbus and awareness can be exchanged within the context of an awareness relationship. Evaluation of awareness also requires:

- a definition of the ranges and meanings of values for focus, nimbus and awareness;
- a means of evaluating focus and nimbus; and
- a means of combining values for focus and nimbus to give an overall awareness value.

These are described in turn.

When describing awareness in the spatial model phrases such as "unaware", "peripherally aware", "more or less aware" and "fully aware" are used. This motivates a quantification of awareness (and also of focus and nimbus) which spans a finite continuous range from "none" to "full". Given the normal properties of real number arithmetic it is appropriate to define an awareness level of 0 to represent no awareness and an awareness level of 1 to represent full awareness. This is the definition adopted by MASSIVE-1. Notions such as "peripherally aware" correspond to intermediate values for awareness.

In the spatial model of interaction focus and nimbus are intended to represent, respectively, an observing object's interests and an observed object's visibility (or audibility, etc.). Examples from the everyday world include:

- human visual focus, which is limited to a particular field of view with a small region of high detail and a larger peripheral region;
- audio focus, which is largely non-directional;
- visual nimbus, which is often (but not always) non-directional (for example one must stand in front of a TV set in order to see the picture); and
- audio nimbus, which is partially directed and which can be radically varied in size (from a whisper to a shout).

For MASSIVE-1 a standard function is defined for evaluating focus and nimbus. This is shown in figure 5.2 on page 59, with the controlling parameters listed in table 5.2 on page 59. It is defined relative to the location and orientation of the object which is calculating it.

The value of the function is sampled at the closest point of the sphere which represents (approximates) the location and extent of the other object to find the value of focus or nimbus on that object. This interpretation of focus and nimbus was adopted as being analogous to observed characteristics of interesting real-world interactions such as the examples of focus and nimbus listed above. It can be seen that the function has both a distance related component (to represent, for example, whispering versus shouting) and an orientation related component (to represent, for example, limited field of view or directed communication). This function is applied in the focus and nimbus calculation stages of awareness negotiation as described above. Each object can change the parameters of table 5.2 on page 59 independently for its focus and its nimbus in each medium. In this way it controls its interaction with the other objects in the world.

Fig.5.2: focus/nimbus function relative to the calculating object's location and orientation and sampled at the peer object

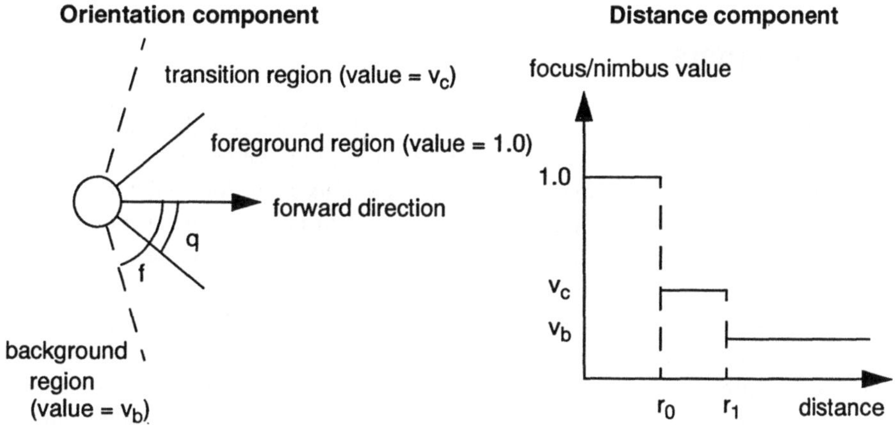

Table 5.2: parameters of the standard focus/nimbus function

name	meaning
θ	conical angle of foreground region
ϕ	conical angle of transition region
v_b	focus/nimbus value of background region
r_0	radius of near-field
r_1	radius of transition region
v_c	value for transition regions

When values for one object's focus and the other object's nimbus have been calculated they must be combined to give a single awareness value. In principle this could be any function which maps two floating point values to one. In MASSIVE-1 multiplication was chosen so that:

- full nimbus (1.0) combines with full focus (1.0) to give full awareness (1.0);
- no nimbus (0.0) combines with any level of focus to give zero awareness - an object cannot observed against its wishes; and
- no focus (0.0) combines with any level of nimbus to give zero awareness - an object can choose to be uninterruptible.

Low levels of focus or nimbus can then represent reluctance to be interrupted or observed, respectively. Philosophically, non-communication is the default, and both objects must cooperate for communication to occur. This would not be the case with other choices of awareness calculation function such as addition (as in [Benford et al., 1993] where either party can force communication).

5.1.3. Adapters

In the spatial model of interaction an adapter may be an object within the space or a metaphor in the user's interface. It transforms the way in which the user or object interacts with the space by modifying one or more of its auras, foci and nimbi. Possible examples of adapters include a megaphone, which increases the size of a user's audio aura and nimbus, and a telescope, which increases the range but reduces the field of view of a user's visual focus.

MASSIVE-1's realisation of adapters is as objects within the virtual world which modify the auras, foci and nimbi of nearby objects (including users' embodiments under the control of client processes). A specialised medium is reserved for the representation and communication of information relating to adapters. Each adapter has an aura in this medium and a medium specific peer interface by which it makes available information about its scope and effect. The effect of an adapter is expressed as a new value for aura size and new sets of parameters for the focus and nimbus functions. Each object which makes use of adapters has its own aura and peer interfaces for the adapter medium. As the object moves about within the space this aura may collide with adapter auras so that the object learns about the adapters that are nearby.

The normal behaviour of an adapted object (such as a user client embodiment) is to change its aura, focus and nimbus parameters to the values specified by the closest adapter. If no adapters are within range then the object returns to its own parameter values. So, as a user moves around a virtual world they enter and leave the regions of influence of adapters and so have their auras, foci and nimbi modified to suit their current environment. Section 5.2.3 describes the adapters which have been implemented.

This concludes the description of MASSIVE-1's realisation of the awareness relationship. The next section describes how this relationship is presented to and may be controlled by a normal user of the system.

5.2. In Use

This section describes how awareness is used and controlled in MASSIVE-1. It draws together and extends material from the first half of chapter 4. The aspects of awareness which are considered are: the influence of awareness on different media; the control which the user has over awareness; the use of adapters; and interaction with communication tools such as the message board.

5.2.1. Presentation

Awareness is calculated and applied to all interaction in every medium (text, audio and graphics). Recall from section 5.1.2 that awareness levels may vary between 0 (no awareness) and 1 (full awareness). The effects of awareness are detailed for each medium in turn.

In the text medium an object is shown in the map view when the user's awareness of it exceeds a threshold value (0.2). The effect of awareness on the presentation of a text message is more complicated - as is detailed in table 4.1 on page 46. Consider an observer's awareness of the message's sender: at very low levels of awareness nothing is observed (the message is not seen by or sent to the observer). At intermediate levels of awareness an observer sees that "something" is said but does not see the message's contents. At higher levels of awareness an observer sees the full message but in brackets (to indicate that the message is not part of a focused exchange). At the highest levels of awareness the full message is displayed. So the text medium is able to indicate directed vs. undirected communication, and also to introduce awareness-based privacy in text communication.

In the audio medium the audio client establishes audio connections only when awareness exceeds a threshold level (0.2). This audio connection is between the users' audio server processes and is in addition to the inter-client audio (management) connection which must exist before awareness calculation can be performed. The audio client controls the playback volume to reflect the user's awareness so that sources heard at low awareness values are quiet, while sources heard at high awareness values are loud. This allows a user to selectively attend to audio sources and groups of sources (e.g. other users) within a crowded and noisy environment.

In the graphical medium an object is rendered when the user's awareness of it exceeds the threshold value (0.2). Objects which are created by the standard world server process may have a number of alternative graphical appearances which are switched between according to awareness level. This can be used for awareness-driven level of detail or to illustrate the operation of awareness negotiation. Note that this is subjective, i.e. specific to each observer according to their own individual awareness of the object.

5.2.2. Focus and Nimbus Control

Awareness and hence interaction are controlled by an object modifying its auras, foci and nimbi. There are three ways in which it may do this: by moving; by changing the parameters which specify focus, nimbus and aura; and by the use of the adapters. The first two of these methods are described in this section while the use of adapters is dealt with in section 5.2.3.

An object's auras, foci and nimbi are normally defined relative to its position and orientation within the virtual world (see figure 5.2 on page 59). As the object moves and turns its foci and nimbi move and turn with it. This reflects the real-world association of the senses to the body - people see in the direction in which they are facing and see and hear things according to where they are. This is the commonest method of manipulating awareness in the spatial model of interaction. One of the aspirations of the model is that people will find this an intuitive, effective and unintrusive means of controlling interaction.

Each user also has a choice between three settings for focus, nimbus and aura. These are "wide", "normal" and "narrow" and provide, respectively, broad undirected interaction (e.g. shout or monitor), mid-range semi-directed interaction (e.g. normal conversation) and local highly directed interaction (e.g. private conversation). These settings are stepped through using a single key press in either the graphical or the text client. In addition users with a graphical client can vary the angle and range of focus and nimbus. This is controlled using the mouse in combination with control keys and is a relatively specialist facility.

The graphical client also provides an optional indication of the user's focus and nimbus. This is represented by a wireframe cuboid which approximates the region of maximum focus and nimbus.

5.2.3. Context-Driven Interaction

Users can indirectly control their interaction using adapters. Adapters are objects which replace a user's auras, foci and nimbi in order to transform the way in which the user interacts and communicates. The two adapters which have been used in MASSIVE-1 are a podium and a conference table. When a person stands on the podium they are given large auras and nimbi so that they can be seen and heard at a greater distance; this permits lecturing and similar patterns of communication. When a person approaches the conference table their auras, foci and nimbi are redefined to encompass the table and its immediate surroundings (including other users at the table) and to restrict interaction further away from the table. This creates a self-contained mutually aware group when participants gather around the conference table.

Adapters in MASSIVE-1 affect individual objects by changing their auras, foci and nimbi and are triggered by proximity. It is simple to create new adapters in MASSIVE-1 which have different effects, i.e. different parameters for aura, focus and nimbus (section 5.1.3 describes the implementation of adapters).

5.2.4. Tool Control

To supplement direct communication MASSIVE-1 includes two communication tools: a message board and a text to speech convertor. The message board displays in the graphical medium the last few messages which it has received in the text medium. Similarly the text to speech convertor monitors the text medium and reproduces these messages in the audio medium using of a freely available (low quality) software speech synthesizer. Each of these tools determines whether a message is directed to it using awareness. This depends on both the message board's focus, which picks out users who are in front of the board and relatively close to it, and the user's nimbus, whatever it may be.

To use either tool a user must approach sufficiently close to it and select a directed nimbus so that the tool has full awareness of the messages which they type. The text to speech convertor (but not the message board) gives visual feedback when it is sufficiently aware of a user.

This completes the description of the implementation and use of awareness in MASSIVE-1. The next section presents observations of and reflection on the use of awareness as it has been implemented.

5.3. Evaluation

The previous two sections have described the implementation of awareness negotiation in MASSIVE-1 and its use and control with respect to normal users. This section comprises observations of and reflections on the effectiveness and shortcomings of the awareness aspects of the system as implemented (chapter 6 considers the network resource requirements of the system). Section 5.3.1 outlines the situations in which MASSIVE-1 has been used and which form the basis for the observations presented in subsequent sections. Section 5.3.2 identifies aspects of the system in use which have been effective. The remaining two sections consider limitations which have emerged with use. These are limitations of personal awareness in section 5.3.3 and problems with navigation and tool control in section 5.3.4. Section 5.4 summarises the main conclusions.

5.3.1. Trials

The major development of MASSIVE-1 was completed in December 1994. Since that time only minor enhancements and bug fixes have been made. The system has been used for approximately 30 formal or semi-formal small-group meetings, plus networked demonstrations and many in-laboratory tests and demonstrations. The context, geographical distribution and approximate numbers of the more formal meetings are listed below.

- Internal weekly group meetings, 4 meetings, over 10 Mbit Ethernet.
- Within the EC-funded COMIC project, 2 meetings, with up to 9 participants spread over 5 sites in 3 countries (the UK, Sweden and Germany), over the Internet.
- Within the EPSRC-funded DEVRL (Distributed Extensible Virtual Reality Laboratory) project [Slater et al., 1996], 4 meetings, between 3 UK Universities (Nottingham, QMW and Lancaster) over SuperJANET.
- Within the BT/JISC-funded Inhabiting The Web (ITW) project [Greenhalgh et al., 1997], 19 meetings, involving 5 UK Universities (Nottingham, Lancaster, UCL, Leeds and Manchester) and British Telecommunications plc, again over SuperJANET.

A typical meeting involved between five and ten participants and lasted for approximately one hour. The participants have normally been computer scientists (often academic or research staff). With the exception of the participants in Germany for the COMIC meetings, all users have had a full complement of text, graphical and audio interfaces. So all participants have been able to communicate via real-time audio, simple graphical gestures and text. Most audio participants used headphones (to reduce echo and disturbance to colleagues), often with an integrated boom microphone. A small number of participants used loudspeakers (where two or more users were sharing a single workstation) with separate desk-top or hand-held microphones. The workstations used were primarily SGI Indys (with unaccelerated and XZ graphics) with some Indigo2 Extremes and High Impacts plus an SGI Onyx RE2 and a Sun SS10ZX.

The trials within the ITW project were held with the express intention of analysing the meetings. The others trials were more open-ended and oriented towards gaining informal experience with the technology. The majority of meetings have been video-recorded from the perspective of one or more participants. For the meetings within the ITW project event log files generated by MASSIVE-1 have also been preserved for a number of the participants and partial records of network traffic have been collected. These are used in chapters 6 and 10 and appendix A.

As well as the evaluations presented in this thesis (in this chapter, chapter 6 and appendix A) MASSIVE-1 has also been evaluated and analysed by Tromp within the COMIC and ITW projects [Tromp, 1995a; Tromp, 1995b; Greenhalgh et al., 1997] and by Bowers, Pycock and O'Brien within the context of the COMIC project [Bowers, O'Brien and Pycock, 1996; Bowers, Pycock and O'Brien, 1996].

5.3.2. Effectiveness

This section describes some of the successes of using MASSIVE-1, which may motivate the future development and potential deployment of CVE technologies.

First, MASSIVE-1 was successful as a tele-conferencing tool. People were able to "meet" (virtually) and communicate. Business was done: where meetings had specific agenda they could be followed and where they had particular goals they were almost always met. It was not a perfect or ideal experience for some of the reasons noted in the following sections and it is beyond the scope of this thesis to formally compare the effectiveness of a meeting held in MASSIVE-1 with other forms of computer mediated communication; the technology is still too immature for this to be a fair comparison. However, there are promising signs both for MASSIVE-1 and for CVEs in general. For example, on one occasion (one of the DEVRL meetings) a MASSIVE-1 tele-conference was chosen in preference to travelling for a face-to-face meeting (and this decision was not regretted after the event).

Second, the flexibility and openness of both the CVE and the spatial model metaphors have demonstrated the ability to support browsing and chance encounters. This was one of the sociological motivations behind the spatial model of interaction. Specifically, people have "bumped into" one another unexpectedly while passing through or exploring MASSIVE-1 worlds. On one occasion a Finnish researcher was discovered exploring the virtual worlds hosted at Nottingham; MASSIVE-1 automatically provided immediate and contextualised audio, graphical and textual communication.

Third, in common with previous research in other areas of computer mediated communication (e.g. [Chapanis, 1975], [Johansen and Bullen, 1984]) it was found (informally) that interactive audio is the single most significant channel for interpersonal communication. By situating interaction within virtual worlds MASSIVE-1 goes beyond current audio conferencing facilities (network-based or otherwise) in terms of flexibility of configuration and (self-) control over participation. For example, a conversation group can easily split into sub-groups for a time by moving to adjacent worlds or just by moving sufficiently far from each other. Individuals can easily and visibly move between groups and sub-groups can merge and reform arbitrarily over time. In each case the CVEs standard metaphor of navigation and situated interaction provides sufficient control and no "mystical" or specialist management intervention is required.

The fourth point of success relates to the general principle of employing space as a resource for interaction. Effective meetings have been held in MASSIVE-1 while employing extremely simple, rather abstract, graphical appearances for both worlds and user embodiments. In terms of stereotypes there are two competing schools of thought concerning visual realism in virtual environments. One argues in favour of photo-realism and accurate humanoid representation of users. The other argues for an impressionistic approach, which concentrates on the basic information to be conveyed (position, orientation, identity, etc.) independent of the normal "real-world" constraints which apply to its presentation. For both pragmatic (performance and authoring) and ideological reasons MASSIVE-1 employs an abstract rather than a realistic approach. For example figure 4.1 on page 43 shows typical user embodiments in a typically simple world. These embodiments convey key information about identity, location, orientation and capabilities but have never yet been described as realistic.

Fifth, Bowers, Pycock and O'Brien, in their analysis of video footage from MASSIVE-1 meetings, find evidence that users make use of their embodiments in ways which are comparable to real-world interaction. Reinforcing the previous point they say ([Bowers, Pycock and O'Brien, 1996]):

> Strikingly, some familiar coordinations of body movement are observed even though such embodiments are very minimal shapes.

Specifically, they observe (also [Bowers, Pycock and O'Brien., 1996]):

> ...the elementary coordination of body movements between participants, the coordination of movements with ongoing speech, the utilisation of the bodies to engage others and initiate talk, amongst other phenomena.

So, at least to some extent, participants are able to appropriate and make constructive use of the facilities for interaction which are made available by the system.

Sixth and finally, by manufacturing appropriate tasks (in the ITW trials) it has been demonstrated that participants are able to make use of the awareness control facilities which are available in MASSIVE-1. Specifically they can make use of the different settings for focus and nimbus. The tasks which have been used to demonstrate this are a team game in which teams within the same space have to avoid being overheard and an identification game in which users explore a space crowded with different audio sources and have to identify and locate individual samples. The team game depends on reducing nimbus to avoid being overheard while the audio game depends on reducing focus to pick out individual sources (recall that audio in this system is not stereo or spatialised and so there are no spatial cues available in the audio signal itself).

These six points show that MASSIVE-1 is useful even in its present (rather unpolished) form. They also give concrete examples of situations which reflect the philosophy and socially motivated approach of the spatial model of interaction. Specifically, there are examples of support for spontaneous and unplanned interaction, the utility of *embodied* interaction, the transfer of uses of space in conversation from the real world to the virtual world and the possibilities of using focus and nimbus to explicitly manage interaction. However, a number of limitations and difficulties have also become apparent. These are described in the following two sections.

5.3.3. Subjective Awareness

The first set of issues relate to awareness, which refers here to the subjective or personal awareness which one user experiences of other participants (rather than the awareness values which exist within the system and which attempt to model and facilitate this "real" awareness). In using MASSIVE-1 three particular problems of awareness have been observed concerning peripheral awareness, engagement between participants and degrees of presence (and absence). These are described in turn.

Peripheral Awareness

First, graphical users experience very limited peripheral awareness. They have no awareness of things which are happening (in the graphical medium) outside of their narrow graphical field of view. The default field of view in MASSIVE-1 is 64 degrees; larger values result in rapidly increasing visual distortion resulting from perspective projection which is particularly noticeable in small windowed displays (current head-mounted displays typically have a still narrower field of view of 40-50 degrees). This can be contrasted with our real-world field of view of approximately 150 degrees. Rapid changes in viewing direction (e.g. glances to left and right) are also much harder in MASSIVE-1 than in the real world, especially with relatively low frame rates on the lower-specification machines. This problem was demonstrated in meetings by the extreme difficulty which users experienced when trying to form a circle because each user was unable to see their immediate neighbours.

The immediate work-around for this problem (one of the extensions made to MASSIVE-1 while it was in use) was to allow graphical users to choose between a range of viewpoints related to their embodiment. The initial version of massive allowed them only to see the world out of their embodiment's eyes. Later versions added the other options listed in section 4.1.1 including "chase" and over-head views. Other solutions to this problem might include: better and faster support for glancing around; the use of audio cues linked to graphical events (as is often the case in the physical world); semi-automated control of viewpoint to capture significant events, or to customise viewpoint according to context (e.g. linked to the operation of adapters); or being able to choose to see out of another user's eyes.

Engagement

The second problem of awareness was a frequent lack of engagement between users attempting to communicate or converse (see [Bowers, Pycock and O'Brien, 1996]). There were various breakdowns in communication, and users were not always confident that they had been heard. Use of back-channels (i.e. conversational responses such as "hmm"s which build confidence in conversation) was infrequent so that users tended to feel as if they were talking into a void. This lack of engagement may be attributed to problems and limitations in both the audio and graphical media. On some occasions, especially in some of the early meetings, the audio quality was erratic and mutual awareness was unreliable causing people to justifiably question whether they had been heard. Apart from these transient problems a more fundamental problem in the audio medium is use of silence suppression which may not reliably differentiate between normal background noise and significant non-verbal sounds (silence suppression is used to reduce the required audio bandwidth and the load imposed on listeners). In the graphical medium a likely problem is that the system is not able to capture or convey detailed information such as gaze direction and other small gestures which play a significant (though often unconscious) role in conversation.

One of the early additions made to the system was a graphical "mouth" which appears when a user is speaking. This allows meeting participants to recognise likely breakdowns in audio communication and to take corrective measures. To a certain extent long term users adjust to the limitations of the system by gaining the confidence to speak into silence, by more explicitly framing the beginning and ends of their utterances and by making more requests for explicit responses in their conversations (e.g. "is that clear?") [Bowers, Pycock and O'Brien, 1996]. Apart from further development and care with all aspects of the audio subsystem, a key area in which this problem might be tackled is that of capturing and conveying gaze direction and other small gestures (see for example [Ohya et al., 1993] and [Thalmann, 1993]). Alternatively live video might be integrated into the graphical medium to provide, for example, user embodiments with video faces as in [Brand, 1987] and [Nakanishi et al., 1996].

Presence

The third problem of awareness was caused by individual participants being able to "leave" their virtual bodies, for example to answer the telephone or to get a cup of coffee. This was not always apparent to other occupants of the virtual world. On occasions one user would appear to deliberately ignore other users. Only with time would it become apparent that the embodiment was currently "unoccupied". This problem has two aspects. The first is that, as with engagement, the normal embodiments are too static to convey (and the system is too limited to capture) the small signs of life which would otherwise indicate a user's virtual presence. The second aspect of the problem is that, as discussed by Bowers, O'Brien and Pycock [1996], virtual reality has tended to assume that users will leave the physical world behind when they enter the virtual world. This is perhaps an unspoken assumption behind aspects of MASSIVE-1's design. What actually happens is that the user remains physically within their normal working environment at the same time as being involved to some extent with the virtual world. Events in the real world (such as the telephone ringing) continue to engage their attention.

In a simple way this problem was addressed in MASSIVE-1 by adding a standard graphical gesture of "sleeping" (the user's embodiment lies down) which a participant can use to indicate that they are not attending to the virtual world. More generally, the same techniques which have been suggested for achieving greater engagement could also be used to determine and represent a user's degree of presence or involvement in the virtual world. It is important that CVE designers recognise the continuing significance of real-world interactions along side activity in the virtual world. Ideally CVEs should provide facilities and tools for communicating and understanding significant real-world events alongside their virtual counterparts.

This completes the description of the problems of awareness which were experienced with MASSIVE-1. The next section considers problems of control and navigation.

5.3.4. Navigation and Control

As noted in section 5.3.2 users were able to use MASSIVE-1 to hold productive meetings. However, in addition to the observed limitations of awareness (above) a number of more "mechanical" aspects of the system presented difficulties for users. These concerned navigation in the virtual world, manipulation of focus and nimbus and control of tools such as the message board. These are dealt with in turn.

Navigation

New users found navigation particularly difficult. Common problems included lack of fine control, falling backwards through portals on entering a new world, becoming lost and disorientated away from the main content of the world and unintentionally using more than the two basic degrees of freedom (fore/aft and left/right) and being unable to re-orient themselves with respect to the ground. Finer grained controls such as moving one's head to look about were also found to be slow and unwieldy. At least in part this may be attributed to the use of standard one and two dimensional input devices only (i.e. keyboard and mouse). The limited field of view already mentioned also contributed to at least some of these difficulties. A further factor was probably the use of a low-level navigational interface expressed solely in terms of relative turns and translations independent of world content.

This suggests two approaches to tackling this problem. The first is the development and employment of more "exotic" interface devices such as 6 degree of freedom trackers and video capture and analysis which might be more suitable than keyboard and mouse for expressing movement and interaction in a three dimensional space. The second possible approach is to provide higher level task-oriented or context sensitive navigational facilities (e.g. moving to a designated object) or to link navigation with the effects of adapters for example. In addition it is important to maintain high frame rates to give accurate movement and timely feedback.

Tool Control

The second problem of control that was found in use concerned the message board tool which has been described in section 4.2.1. The message board takes text messages which it observes above a certain (high) level of awareness and displays them in the graphical medium. The message board was created with a relatively small default focus and the standard options for user nimbi were created so that only the non-default "narrow" setting could be used to trigger the message board. To some extent displaying messages on the board was made deliberately difficult to avoid accidentally over-writing important information. However in normal use very few users were able to write on the message board even when they wanted to. Writing on the board requires a combination of correct choice of nimbus plus accurate spatial positioning. This is complicated by the lack of feedback of awareness level in the graphical medium.

The limited solution adopted for this problem in the ITW meetings was to prepare additional teaching material and to organise a training session on the use of the message board. This was an effective solution but is neither general nor elegant. This problem suggests that users are not sufficiently supported in using or understanding the spatial model as it affects their interaction in MASSIVE-1. For example, it demonstrates the need for more explicit information in all media about critical awareness values, as well as indications of why values are as they are and how they might be changed. In the case of the message board the user needed to know from the graphical medium when they were engaging the message board and needed to distinguish between the need to get closer to the message board (more within its focus) and the need to increase the "power" of their own nimbus. Ideally both of these corrective actions could be directly presented as options to the user.

Interaction Control

The final problem concerned the employment of the spatial model in normal interaction. When the trials began the default settings for focus and nimbus were "normal", i.e. moderate in extent and directed to emphasise face-on interaction. However the default setting was soon changed to "wide", i.e. large in extent and undirected. In either case participants rarely changed their focus/nimbus setting in normal use.

In part this may be due to the nature of the trials, all of which involved small numbers of participants (up to 10) in a single group. In this context participants normally expected to be able to talk to everyone else in the same environment and there was almost never more that a single active conversation. Consequently the "normal" setting of focus and nimbus was found to disrupt interaction rather than facilitate it. For example people would be unexpectedly quiet, or outside of audio range altogether. With the "wide" setting interaction management degenerated to a single interactional group per world with little or no spatial dependence except in specialised tasks (such as using the message board). One of the suggested strengths of the spatial model of interaction is in dealing with interaction in large and highly populated environments but it has not been possible to test this in MASSIVE-1 (because it cannot support larger numbers of mutually aware users on the machines available). In smaller well-defined groups the affordances and facilities of the spatial model do not typically come into play. Because of this specific tasks were created which required the flexibility of the spatial model even with relatively small numbers of participants (see section 5.3.2). Practically testing the spatial model's support for large-scale interaction will have to wait for a system which supports much larger numbers of participants.

It is likely that users' apparent reluctance to actively employ the spatial model also reflects the lack of feedback which they receive about focus, nimbus and awareness. Furthermore the options to change focus and nimbus are not visible in the interface, but rely on users recalling specific key presses. Future systems which build on the spatial model should give consideration to the representation of focus, nimbus and awareness to the user, both within the virtual environment and the interface.

A final reason for the under-utilisation of the spatial model may be that the fixed functions representing focus and nimbus are too inflexible and do not take into account the context in which interaction occurs. For example, in the everyday world someone would (consciously or unconsciously) adjust the volume of their voice to compensate for the space which they are in and the people that they wish to address. MASSIVE-1 has no comparable automatic mechanism for tuning focus and nimbus. Similarly, someone might concentrate on the *nearest* conversation, wherever it might be. Again, MASSIVE-1 cannot directly express this kind of focus. Further development and evaluation is needed in this area to refine descriptions of focus and nimbus to achieve a fuller range of real-world interaction styles without requiring explicit user intervention. This area is also addressed by the extensions to the spatial model which are proposed in chapter 7 which allow the spatial context in which interaction occurs to affect awareness (e.g. being in a closed room vs. being in a open park).

This concludes the observations of and reflections on support for awareness and interaction in MASSIVE-1. The next and final section of this chapter summarises the main points and draws out general conclusions for CVEs and the spatial model approach to awareness.

5.4. Summary and Conclusions

This chapter has described the implementation of direct awareness relationships in MASSIVE-1 and the way in which awareness is used and controlled in normal interaction. The previous section presented an evaluation of this from the perspective of the user, in the form of observations and reflections based on approximately 30 meetings held using the system. This section summarises the main points of the evaluation and highlights its conclusions. This is done in three stages: first, with respect to CVEs; second with respect to CSCW; and third with respect to the spatial model of interaction.

5.4.1. CVEs

With respect to CVEs, experience with MASSIVE-1 has shown that they can be a useful and effective technology to support synchronous collaboration and tele-conferencing. However it has also highlighted four shortcomings of current CVE systems which are listed below.

- Support for peripheral awareness. It was found that viewing the virtual world (graphically) from "out of the user's eyes" dramatically limited their peripheral awareness, that is their general awareness of surrounding objects and activity. So CVE designers must consider carefully the form and degree of coupling which is enforced between a user's embodiment and their perception of the virtual world. For example, multiple camera views (such as those added for MASSIVE-1) give the user more flexible ways of viewing the world but make it harder for other users to reason about what the user can see.
- Ease of navigation. Navigation was found to be awkward, especially for new users. This is a fundamental issue for CVEs because space and the use of space lie at the heart of the CVE approach. This is also an issue for all 3D interaction including visualisation and single-user VR interfaces.
- Flexible notions of awareness. Current CVEs do not provide explicit support for social factors in awareness. MASSIVE-1 has demonstrated that flexible notions of awareness can be deployed and used in a CVE to complete tasks which would otherwise be impossible (e.g. forming private subgroups, working in environments with audio "clutter").
- Support for concurrent involvement in real-world activities. The VR approach has tended to assume that users will effectively leave the physical world behind

when they enter the virtual world. What actually happens, especially with desktop rather than immersive systems, is that the user remains physically within their normal working environment at the same time as being involved to some extent with the virtual world. CVE designers need to recognise the continuing significance of real-world interactions alongside activity in the virtual world ([Bowers, O'Brien and Pycock, 1996]).

5.4.2. CSCW

With respect to CSCW MASSIVE-1 has demonstrated on a number of occasions that it (and the underlying model) can actively support spontaneous and unplanned interaction. This also reflects the nature of the aura relationship which is dealt with in chapter 6 (though from a more network-oriented perspective). Interaction in the everyday world can exhibit great flexibility and informality (see section 3.1). It is important that CSCW and distributed system designers do not assume a tacit and naive over-formal model of action and interaction which may unnecessarily limit the usefulness of such a system (see [Bowers et al., 1995] as an example of the conflicts that can arise between formalised models of working practice and reality). CVEs in general, and the spatial model of interaction in particular, are two approaches to supporting collaboration which aim to actively support informal and unplanned interaction and collaboration.

5.4.3. The Spatial Model

Experience with MASSIVE-1 has produced examples of situations and activities which bear out some of the motivations behind the spatial model of interaction. These include supporting unplanned and spontaneous interaction, the utility of embodiment within a spatial frame to convey information and to support negotiation of interaction, the transfer of real-world spatial actions to the virtual world and the necessity of both focus and nimbus in different situations. These observations encourage further development and investigation of the spatial model of interaction and similar approaches. However, use of MASSIVE-1 has also demonstrated a number of shortcomings in the current realisation of the model. These are listed below.

- Feedback and the malleability of the model. Feedback about and control of the spatial model has proved to be inadequate in the current system. This has created problems with the use of awareness-controlled tools and contributed to the general neglect of focus and nimbus facilities. Future systems should give users direct and visible feedback about current awareness relationships and the influences of focus and nimbus.
- Sensitivity to context. Basing awareness solely on fixed focus and nimbus functions defined over an abstract space has proved inadequate and at times frustrating. This problem reflects the neglect of two aspects of interaction: the effects of

spatial context; and accommodation of (or adaptation to) interaction partners. The former aspect is addressed by the extensions to the model proposed in chapter 7 and prototyped in MASSIVE-2 (chapters 8 to 10). The latter aspect of interaction requires a more flexible, less rigidly spatial approach to specifying focus and nimbus.

- Scalability. It has not been possible using MASSIVE-1 to test the model in large or highly-populated worlds because of the limited capabilities of the system. One of the goals of MASSIVE-2 is to support larger numbers of users and hence to push the model in this respect.
- The difficulty of supporting rich interaction when compared to face to face interaction. The straightforward implementation of the spatial model of interaction which is embodied in MASSIVE-1 has not (magically?) given rise to the richness and naturalness of interaction which might have been hoped for. Face to face interaction in the physical world does not "just happen". It is not clear at the present time whether the observed limitations of MASSIVE-1 are because other mechanisms involved in negotiating interaction have been neglected (e.g. gaze direction, small gestures, sub-vocal audio cues) or because the use of space and focus and nimbus is not sufficiently refined.

This concludes chapter 5 which has presented a description and evaluation of awareness negotiation as realised in MASSIVE-1. The emphasis in this first part of the thesis is on *direct* relationships - the awareness and aura relationships of MASSIVE-1 and the spatial model of interaction. This chapter has adopted a social perspective, particularly with regard to evaluation. The next chapter brings a more computational perspective to its consideration of the aura relationship and the concept of spatial trading. The focus there is more explicitly on scalability, especially with respect to network communication.

6.
Spatial Trading

Section 4.4 of chapter 4 described how the spatial model of interaction was realised in MASSIVE-1 in terms of two key relationships: the awareness relationship, which links two interacting objects, and the aura relationship, which links an object to a world. Chapter 5 has considered the awareness relationship and the way it which it is used to realise and negotiate direct awareness between objects (and participants). The evaluation of that chapter focused on the way in which this was made available to and used by normal system users. This chapter focuses on the communication management employed in MASSIVE-2 and its implications for the scalability of unicast peer-to-peer CVEs. This is presented in terms of "spatial trading", one of the main contributions of this thesis, which is the formalisation of aura management in a distributed systems context. Section 6.1 describes how spatial trading is implemented in MASSIVE-1 in terms of the aura relationship while section 6.2 describes how spatial trading is used and managed. Section 6.3 evaluates the scalability of spatial trading and unicast-based peer-to-peer networked CVEs with respect to total network bandwidth requirements. A networking perspective is adopted for the evaluation in this chapter to complement the more sociological perspective of the previous chapter. Finally, section 6.4 summarises these results and sets the context for part II of this thesis. However first of all spatial trading must be defined more formally.

In the spatial model an aura is "a subspace which effectively bounds the presence of an object within a given medium and which acts as an enabler of potential interaction" [Benford and Fahlén, 1993]. So in the spatial model objects with overlapping auras should be enabled to communicate. This can be conceptualised as a brokering or trading process in which objects with compatible and overlapping auras are introduced to one another. This is formalised here as the concept of "spatial trading" (as reported in [Greenhalgh and Benford, 1995]), which combines aspects of attribute-based naming services as in ANSA [ANSA, 1989] and ODP [ITU-T, 1995] with the virtual reality technique of collision detection.

In trading terms an aura is an offer of service and a request for service combined. The attributes which characterise an aura are: world; medium; location with the world; and size and shape. Figure 6.1 on page 76 illustrates the concept of spatial trading. Each client keeps the spatial trader up to date with details of its aura (which combines offer and request). The spatial trader uses collision detection techniques to incrementally identify matching offers and requests and informs the corresponding clients. Clients then act on this information, for example, to establish direct channels of communication as in MASSIVE-1 (which establishes an awareness relationship as described in chapter 5).

Fig.6.1: the concept of spatial trading

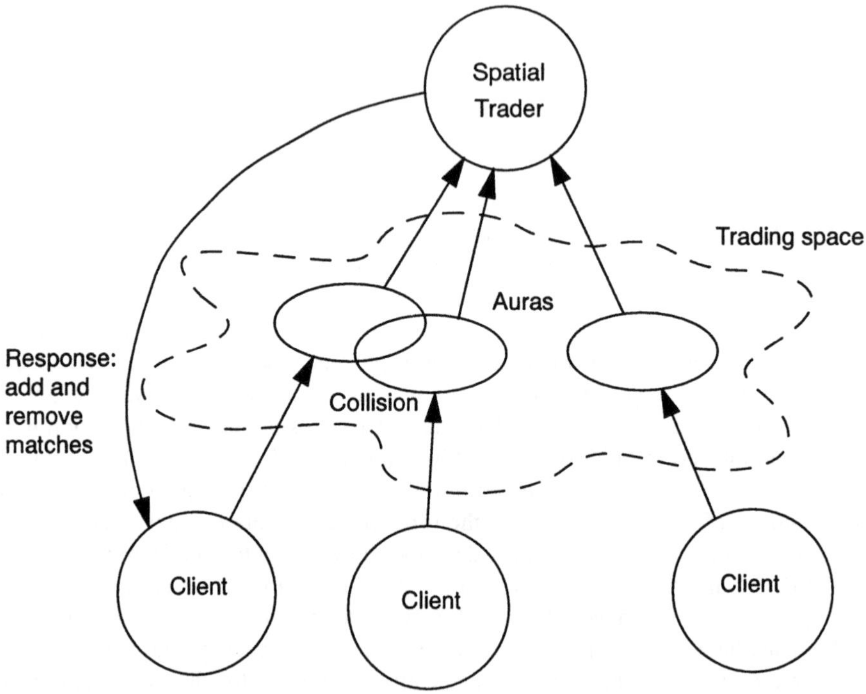

There are four features of spatial trading which distinguish it from normal attribute-based trading (as in the ODP model, for example). These are listed below.

- The offer and request persist together for as long as the object exists. In contrast, the normal behaviour of a trading service is that, while offers may be long-lived, requests are transient and return a single response. Normally requests have no ongoing effect on or representation within the trading service.
- The criteria for matching offers and requests include a notion of distance or proximity. In attribute-based trading the matching criteria are more often discrete pass/fail tests such as matching a text string.
- A request always expects to learn of all matching offers. Attribute-based traders typically support this as an option but it is more common to search for a single matching offer (e.g. representing a provider of a particular service).
- Over their lifetimes offers and requests may change many times and typically each change will be small (e.g. as an object moves around a world). With attribute-based trading offers and requests are normally fixed when they are cre-

ated and there is no efficient mechanism for incrementally changing them (e.g. avoiding renotification of offers which are already known).

The formulation of spatial trading is one of the main outcomes of this thesis. The next section describes how it is implemented in MASSIVE-1.

6.1. Implementation

An implementation of aura-based spatial trading requires:

- a spatial trader (or "aura manager" in figure 4.5 on page 52 and figure 5.1 on page 56) which monitors auras and performs introductions;
- a protocol by which an object can communicate with the spatial trader, specifying its aura attributes to the spatial trader and receiving responses from it.

In MASSIVE-1's distribution model the spatial trading protocol is realised as a connection between appropriate interface types: an object or process which depends on spatial trading creates an instance of the *aura* interface type and connects it to the spatial trader's (null) interface (see figure 6.2 on page 78). This connection forms the context in which an aura can be efficiently and incrementally updated and which can persist for as long as required (but not longer - failure of the object will terminate the connection, signalling withdrawal of the aura).

Because the spatial model defines worlds as disjoint spaces there is no possibility of an aura in one world colliding with an aura in another world. Consequently it is simplest to organise spatial trading on a per-world basis. In MASSIVE-1 a single spatial trader may be responsible for one or more worlds. Other spatial traders may be running at the same time (on the same or other machines) and have responsibility for other worlds. In as much as a world is explicitly represented in MASSIVE-1 it is represented by the spatial trader (or aura manager) responsible for that world, i.e. in MASSIVE-1 the only objective (universally agreed) information about a world is the identities and locations of the auras within that world. All other information, e.g. object appearances, are communicated directly between the objects concerned and may be different for each observer. The spatial trader has no information about the objects "behind" the auras: it has no knowledge of the contents of individual media such as geometry, sound or text. So the relationship between an object and a world is defined entirely by the object's auras.

The remainder of this section gives details of the spatial trading protocol, which is the realisation of the aura relationship (section 6.1.1), and explains how spatial traders are located and used (section 6.1.2).

Fig.6.2: a spatial trader with a single client

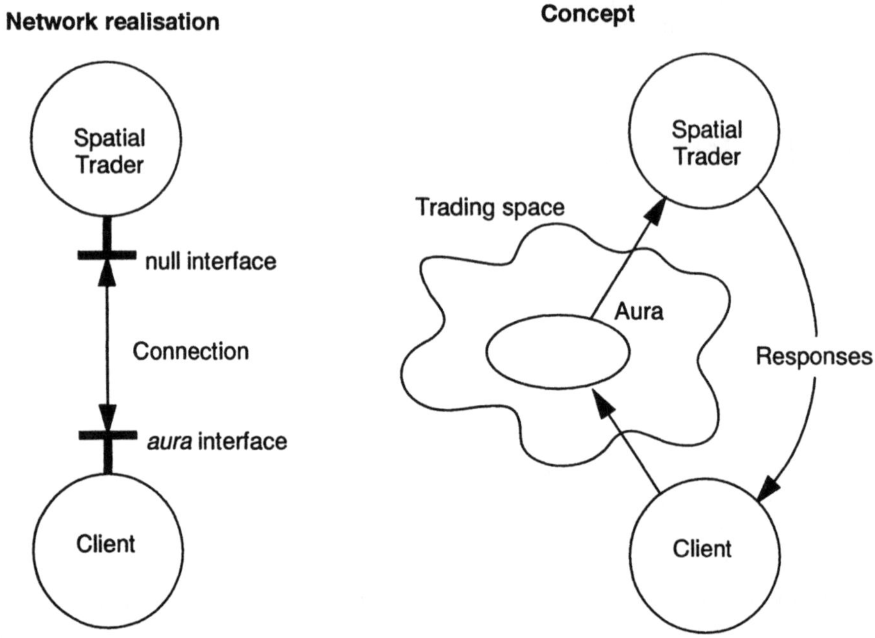

6.1.1. Aura Relationship

This section describes the aura relationship, i.e. the spatial trading protocol. Consider the example in figure 6.2 on page 78 which shows a single client process and a single spatial trader process. The client process has an *aura* interface which represents a single medium-specific aura. The client establishes a connection between this interface and the spatial trader's interface (which has a null type and does not define any attributes or entry points). The *aura* interface type is shown in table 6.1 on page 79.

The first six attributes describe the aura:

- *identity* explicitly identifies the object which owns the aura and is needed to avoid race conditions which might otherwise occur in the implementation;
- *world* identifies the virtual world using a readable string (the world name);
- *medium* specifies the name of the medium in question and implies a particular type of peer interface although this is not checked (the spatial trader just matches text strings, so that if a new kind of object joins the system which uses a new

Table 6.1: the *aura* interface type

Purpose	Name	Kind	Type
Describe aura	identity	attribute	identity_t
	world	attribute	string
	medium	attribute	string
	position	attribute	float[3]
	radius	attribute	float
	type	attribute	aura_type
Offer details	interface	attribute	interface_address
Response	collision	stream	collision_t *(match/end match)*
Management	add manager	action	interface_address, returns int
	remove manager	stream	void

medium then the spatial trader can dynamically add that new medium to the running system);

- *position* and *radius* specify a spherical aura in the three-dimensional world space; and
- *type* distinguishes between the normal active auras associated with users and passive auras which are associated with background objects - passive auras should not be introduced to one another since neither party wants to obtain information (this corresponds to an "offer-only" aura and avoids generating a large number of unused collisions and notifications within the system and the spatial trader).

The information which is exchanged when two compatible auras collide is the *interface* attribute which specifies the network address of an interface which can be used to communicate directly with the object concerned. In MASSIVE-1 this will be the appropriate medium-specific *peer* interface.

The *collision* stream is used by the spatial trader to notify a client when its aura collides with and separates from other compatible auras. This is the end result of the trading process. Note that separation of auras is also notified to the client. This prevents the number of active matches increasing indefinitely and avoids race conditions which would result if this decision were delegated to the individual clients involved.

The *add manager* and *remove manager* entries are used to manage multiple spatial traders and are dealt with in the next section. To summarise the main part of the aura relationship, a client with an aura creates a corresponding *aura* interface and connects in to spatial trader. The client sets the *identity*, *interface* and *aura type* attributes and then uses the *world*, *medium*, *position* and *radius* attributes to communicate with the spatial trader. The spatial trader returns *collision* notifications which include the *interface* attribute from the colliding aura's interface so that the clients can establish direct communication.

6.1.2. Managing Multiple Spatial Traders

It was noted in the previous section that a number of spatial traders may be running concurrently, each handling one or more worlds. The last two elements of the aura interface, *add manager* and *remove manager*, support a referral facility by which a spatial trader can pass an aura on to another spatial trader. This is used to move auras to the appropriate spatial trader for their current world. Transfer of an aura may occur in two situations:

- when a client starts up it connects to a default spatial trader on a well-known address which is not necessarily responsible for the aura's world; and
- when a client moves from one world to another (e.g. via a portal) it will specify the new destination world which may be handled by a different spatial trader.

In each case the current spatial trader uses preconfigured information about worlds managed by other spatial traders to transfer the aura to the appropriate spatial trader. A spatial trader uses the *add manager* action (RPC) to redirect the client to the destination spatial trader; it then uses the *remove manager* stream event (asynchronous message send) to tell the client to disconnect from the referring spatial trader. Figure 6.3 on page 81 illustrates this referral process for a new client process.

It would be straight-forward to use global locators such as URLs to identify spatial traders for particular worlds. This would allow the evolution of a widely distributed and loosely coupled virtual universe (linked via portals) directly comparable to the World Wide Web.

This section has described the aura relationship in MASSIVE-1 which is realised in terms of spatial trading. It has also given details of the aura protocol and of the interaction between multiple spatial traders. The next section describes how spatial trading is seen and controlled by normal users of the system.

6.2. Use

This section describes briefly how aura and spatial trading are used and controlled in MASSIVE-1 as viewed by a normal user.

As in the spatial model, aura collision is a prerequisite of interaction in MASSIVE-1. Without aura collision there can be no interaction and no awareness between objects. Auras are specific to a single world and medium and so all direct interaction occurs between objects which are in the same world and which support the same medium (although a single object can also have other auras in other media).

Fig.6.3: the process of referral between spatial traders for a new client

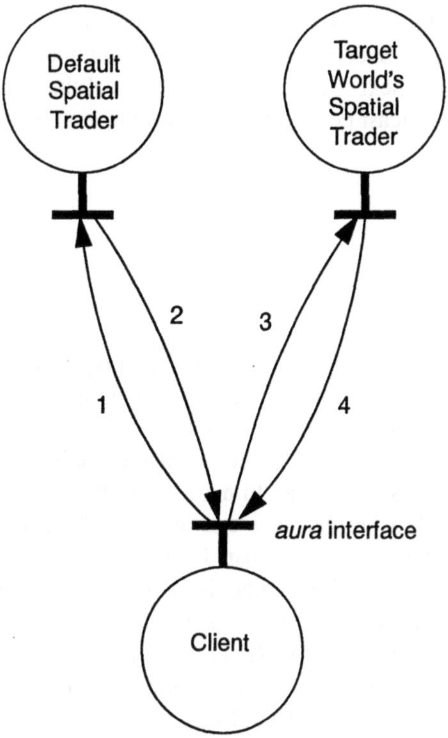

1. An object connects to its default spatial trader and describes its aura, specifying its desired world.

2. The default spatial trader refers the object to the designated spatial trader for that world and tells it to disconnect from the default trader.

3. The object connects to the new spatial trader and describes its aura to it.

4. The new spatial trader, which handles that world, tests for aura collisions and informs the client as they occur.

Until auras have collided, an awareness relationship will not be established and so awareness cannot be calculated. So a user is unable to perceive (see, hear, etc.) any object which is out of aura range. Those objects which are within aura range may be further filtered before their presentation to the user based on negotiated awareness levels, as described in chapter 5. Hence aura defines the maximum scope of a user's awareness within the virtual world. An aura should normally be at least as large as the object itself (which might be a room or a building). This ensures that any objects which are *within* an object will be within its aura range and so made aware of it. For an active object such as a user, an aura must also be at least as large as their focus or there will be sudden jumps in awareness when an object eventually comes into aura range.

An object's auras are continually updated as it moves around the virtual world and as it jumps between worlds. This updating occurs automatically as users navigate using the facilities of the text and graphical clients (described in chapter 4, section 4.1). In addition the size of a user's aura can be controlled directly together with focus and nimbus. The three preset interaction modes ("wide", "normal" and "narrow") provided by the text and graphical clients each have their own appropriate aura size. The graphical client facilities which vary focus and nimbus also adjust aura size.

Finally, each adapter object specifies an appropriate size for aura in addition to specifying parameter values for focus and nimbus.

This section has described how aura is used and controlled in MASSIVE-1. The use of aura and spatial trading is straightforward because the system has been designed "from the ground up" around these concepts: no inter-user communication ever occurs without aura playing its part. The next section considers the usefulness of aura and the scalability of CVEs based on unicast peer-to-peer communication (such as MASSIVE-1).

6.3. Evaluation

The evaluation in this chapter adopts a computational and networking perspective to complement the more sociological perspective of chapter 5. The last two sections have described how spatial trading is implemented and used in MASSIVE-1. This section now considers the potential *scalability* of a CVE based on spatial trading and of unicast-based peer-to-peer networked CVEs in general. In many ways this section is a prelude to the evaluation of chapter 10. It is preserved here to maintain the historical structure of the work and to motivate the transition to part II of this thesis. This evaluation is also somewhat less extensive and ambitious than the analysis of section 10.3 and provides a gentler introduction to the issues under consideration.

This section analyses the potential scalability of MASSIVE-1, i.e. the number of simultaneous users which it may support. It does this by developing a predictive model of the total network bandwidth requirements of MASSIVE-1 for varying numbers of users. This traffic model is then used to estimate the network bandwidth requirements for different user populations and equivalently to estimate the numbers of simultaneous users which could be supported by given networks. The model is also used to explore the effect of spatial trading and the potential utility of network multicasting (rather than the unicast communication employed in MASSIVE-1). This analysis assumes that the total network bandwidth is the limiting factor determining scalability, an assumption which is partially relaxed in chapter 10.

It has not been possible to use MASSIVE-1 with very large numbers of simultaneous users: the largest meeting to date involved 10 participants. However the traffic model described here builds on the experience to date, combined with detailed analysis and understanding of the protocols and operation of MASSIVE-1 to extrapolate to larger scale usage. The traffic model combines:

- a simple model of user behaviour based on the data in appendix A which has been derived from analysis of the MASSIVE-1 trials listed in section 5.3.1, especially the ITW trials;
- a detailed study of the various protocols used (as described in this chapter and the previous one) with regard to the network messages associated with key user-level events; and

- a number of assumptions about the kinds of interaction which may take place.

The modelling of user behaviour characterises the user activity observed in small scale trials. Strictly, this is specific to small-group teleconferencing with the particular users involved - other users in other applications and settings may behave very differently. However it provides a starting point in an otherwise undefined area.

User actions such as moving and speaking have direct implications for the system which must propagate this information to other participants. By analysing the protocols used, it is possible to predict network traffic requirements as a function of user activity. When combined with knowledge of system handling of additional users and with assumptions about general patterns of activity, it is possible to extend this prediction of network traffic requirements to address arbitrary numbers of simultaneous users. This is the approach adopted here (section 10.3 describes this modelling process more generally).

Section 6.3.1 describes the model of user behaviour employed. Section 6.3.2 lists the key data exchanges found in MASSIVE-1 which depend on user activity. In section 6.3.3 this data is combined to give the full traffic model which shows how total bandwidth requirements are expected to vary with the number of simultaneous users and the use of spatial trading. Section 6.3.4 reflects on the validity and generality of this model. The last two sections consider the implications of this model for spatial trading (in section 6.3.5) and for unicast peer-to-peer networked CVEs in general (in section 6.3.6).

6.3.1. User Model

The key elements of the model are listed below. First there are a number of assumptions or restrictions on the use of the system.

- All users have a full complement of text, graphical and audio clients.
- Users have standard embodiments.

Second, there are a number of simplifying observations of the typical characteristics of MASSIVE-1.

- The most significant user events are moving and speaking (text messages and gestures are ignored).
- The average frame rate is 6 Hz (observed in use on a Sun 10ZX, and adequate for normal use).
- The contribution of background objects is much less than that of users and can be ignored (from preliminary studies a background object in MASSIVE-1 generates roughly one tenth the traffic of a normal participant).

Third, there are two observations of "typical" user activity based on analysis of small-scale trials (as described in appendix A).

- Users move 20% of the time (see appendix A, section A.2.1).
- Users speak (or rather send network audio data) 25% of the time (appendix A, section A.3.1).

Finally, there are two more general assumptions about the way in which the system might be used and people interact and behave.

- Users move between worlds or groups of users so that they change the peers with which they interact once per minute (this value is somewhat higher than in the trial meetings, which was about 5 minutes, but remains a relatively insignificant part of the total).
- All users interact with (on average) M other users as a result of spatial trading. For example, users might form variable and changing groups with an average of $M + 1$ participants.

6.3.2. Application Model

In MASSIVE-1 the main causes of network traffic are:

- coordinating master and slave user clients;
- updating the spatial trader (due to movement, world transition or aura adaptation);
- establishing new associations with other users and objects upon aura collision; and
- interacting with other users and objects while in aura range.

Of these four items the first two are independent of the number of other participants using the system. The third item, establishing new associations, depends on the rate at which groups or associates change. The fourth, interaction, depends on the number of other users and objects within aura range at any time (which is denoted by M).

Table 6.2: network traffic resulting from key events

	Movement (kbyte/step)	Audio (kbyte/sec)	New peer (kbyte/peer)
Per user	1.2	0	2.1
Per peer per user	2.1	8.3	13.2

The three key user-level events or activities which generate associated network traffic are: user movement; speech; and the arrival of a new interaction partner (i.e. an aura collision and subsequent data exchange). Table 6.2 on page 84 shows the basic traffic generated by each of these events independent of any particular assumptions about user behaviour; each has a component which is per user only (the upper row), and a component which depends on the number of peers which each participant currently has (the lower row).

6.3.3. Traffic Model

The user model in section 6.3.1 indicates how often each of the events in table 6.2 on page 84 can be expected to occur. Combining this information gives average network bandwidths for each of these activities as shown in table 6.3 on page 85. The user model applied assumes an average movement rate of 1.2Hz (taking into account frame-rate and stationary periods), a speaking rate of 25% and a peer change rate of once in 60 seconds. These combine with the values for movement, audio and new peers, respectively.

Table 6.3: average network bandwidths

	Movement (kbyte/sec)	**Audio (kbyte/sec)**	**New peer (kbyte/sec)**	**Total (kbyte/sec)**
Per user	1.4	0	<0.1	1.4
Per peer per user	2.5	2.1	0.2	4.8

The final column of table 6.3 on page 85 gives the total expected network bandwidth (in kbyte/sec) as a function of the total number of participants, N, and the average number of other participants in aura range, M:

$$B = N(4.8M + 1.4) \quad \text{(Equation 6-1)}$$

For constant group size (i.e. constant M) this will be proportional to N, the total number of participants. On the other hand, if all users are in aura range (i.e. effectively a single group) then $M = N - 1$ and the maximum possible bandwidth is obtained:

$$B_{max} = 4.8N^2 - 3.4N \quad \text{(Equation 6-2)}$$

This is equivalent to using neither aura nor multiple worlds to limit interaction. Figure 6.4 on page 86 shows the resulting total bandwidth plotted against the number of simultaneous participants for different group sizes (i.e. different values of M). Figure 6.5 on page 86 shows the tradeoffs between total number of participants and average group size for a range of total network bandwidths.

Fig.6.4: total network bandwidth against number of participants for a range of group sizes

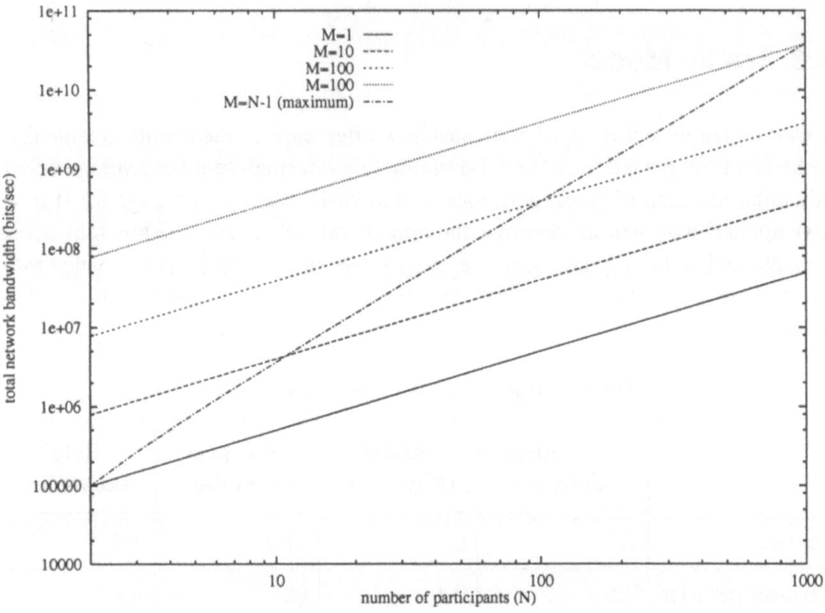

Fig.6.5: trade-off between number of participants and group size for a range of total network bandwidths

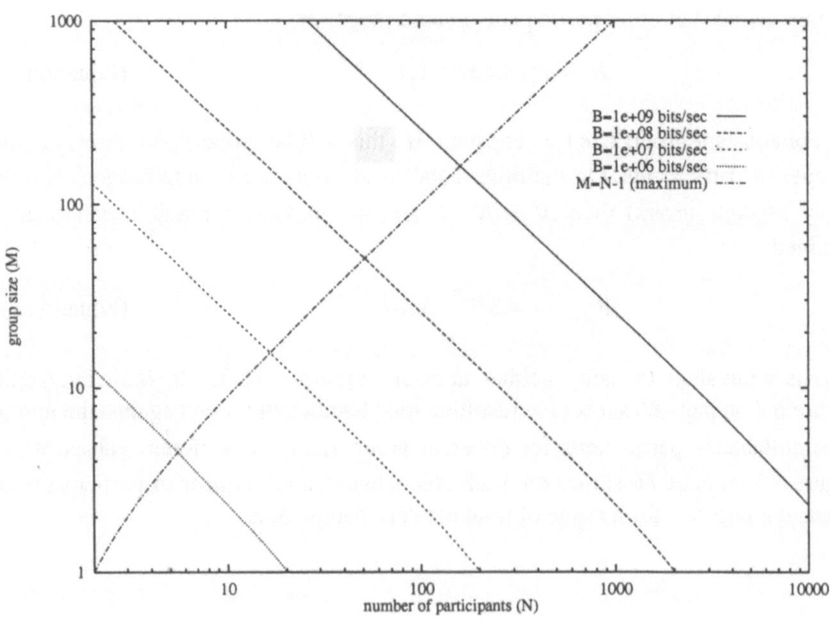

6.3.4. Reflections

Having presented a simple model of total network traffic for MASSIVE-1 it is appropriate to reflect on its limitations before using it to explore implications for the use of spatial trading and the scalability of unicast-based peer-to-peer CVEs.

Observed network traffic has been analysed for one machine for six of the ITW meetings (see section 5.3.1) to provide a qualitative check on the values in the model. It was not possible to track and identify the exact event-related traffic used in the model. The observed average bandwidths are shown in table 6.4 on page 87.

Table 6.4: observed average network bandwidths for MASSIVE-1

Traffic	Per user	Per peer per user	Model
Audio	-	1600 bytes/s	2100 bytes/s
Movement	-	1300 bytes/s	2500 bytes/s
Spatial trading	100 bytes/s	-	part of movement and new peer - 1400 byte/s
Local trading	10 bytes/s	-	neglected
World server	300 bytes/s	-	neglected (background objects)

The most important figures, for audio and movement, are comparable with the model values. The low figure for movement probably reflects the use of a less capable machine for these measurements (supporting a lower than average frame-rate). The communication with world servers is found to be relatively small, justifying the neglect of background objects in this application. It was not possible to track the inter-client communication which is a significant part of the per-user components of the traffic model. Altogether, these measurements have the expected form of per-user and per-peer components and are consistent with the model presented.

Having touched on the experimental validity of the model it is worth revisiting the assumptions that underlie the model. These assumptions limit the accuracy and appropriateness of the model in different situations and scenarios and should be borne in mind when applying these results. The main reservations concerning the applicability of the model are listed below. Some are rather minor, while others are significant for some applications such as data visualisation.

- MASSIVE-1 uses unicast-based peer-to-peer networking; the model and inferences are specific to this approach.
- The impact of background objects has been ignored. This may be appropriate in a tele-conferencing application but would not be appropriate in a database visuali-

sation which might have thousands or millions of non-user objects. To apply the model in this kind of application requires that objects be accounted for (at least) in terms of equivalent numbers of participants or influence on effective group size.

- Particular rates of change and levels of complexity of peers (e.g. geometry complexity) have been assumed. With these assumptions traffic due to the establishment of new peer associations is small (less than 5% of the total). However a combination of rapid change and much greater peer complexity could dramatically increase this traffic component and change the pattern of total system bandwidth requirements.
- Different systems will use different protocols giving rise to different amounts of traffic for each key event. MASSIVE-1 is rather inefficient (e.g. duplicating movement messages for each medium). However no one class of event is *disproportionately* inefficient so the same pattern of behaviour should be observed in other systems, though probably for larger numbers of participants or for larger group sizes.
- Different levels of user activity will result in different bandwidth requirements for each user (e.g. there will be more audio traffic if individuals spend less time silent). Again, however, the overall form of the results should be the same.

So this traffic model remains representative in form of any CVE system based on unicast peer-to-peer communication in conjunction with either spatial trading or multiple worlds (see below).

6.3.5. Spatial Trading

The model of expected network traffic demonstrates the potential of auras and spatial trading to reduce network requirements and/or increase the potential number of participants when compared to a system which does not use spatial trading. Most significantly, the use of auras reduces the total network bandwidth from $O(N^2)$ (equation 2) to $O(NM)$ (equation 1). The management traffic overhead due to interacting with the spatial trader is quite small (it is one component of the per user traffic in table 6.2 and table 6.3 on page 85) and is $O(N)$.

Using auras to scope (and limit) interaction in this way has the same effect on network bandwidth as enforcing a distribution of participants between a number of different virtual worlds. However using auras and spatial trading has significant advantages over splitting participants just according to world. These advantages of spatial trading are listed below.

- It naturally includes division by world as a special case.
- It allows gradual, natural and visible transitions between groups within the same world, as opposed to sudden and discrete jumps between worlds.

- It allows different media to be treated in different ways (e.g. large visual aura and small audio aura).
- It avoids the need for invasive system intervention such as barring access to busy worlds.
- It supports flexible control and graceful degradation through interactive or automatic modification of aura, for example shrinking auras to reduce system and network load in a busy region.

The use of auras and spatial trading as the basic enabler of communication has been one of the most successful and interesting aspects of MASSIVE-1.

6.3.6. Unicast CVEs

Even with spatial trading to manage the number of active peers it is clear from this traffic model that peer-to-peer interaction is the dominant component of network bandwidth, even with relatively small numbers of active peers. This is the "per peer per user" component in the model, which is $O(NM)$. In a CVE there are elements of interaction which are specific to a single pair of peers; this is particularly clear in a system based on mutually negotiated awareness as MASSIVE-1 is. However the basis of a CVE is normally a shared and primarily objective virtual world: participants can agree about the locations and properties of objects and embodiments. This objectivity and agreement are significant for facilitating cooperation. Consequently much, if not all, of the information being communicated between participants' processes is objective in character, i.e. observer independent.

The unicast-based networking employed in MASSIVE-1 means that each network packet (each message or block of information) has to be delivered to a single destination. So to send the same information to a number of destinations (e.g. a number of peers) requires one packet for each destination. These packets often travel over the same network links and segments for some or all of their journeys. It is this duplication of packets which gives rise to the $O(NM)$ limiting term in the network traffic model because N users must each send M copies of each message.

However network multicasting allows the same packet to be delivered to many destinations. With networking technologies that use a shared physical medium (e.g. Ethernet and FDDI) the same packet can be directly received by any number of machines on that network segment. The addition of multicast routers allows a single packet to be distributed to many destinations over wide area and inter-networks (see figure 2.1 on page 28). Using multicast routers defers duplication of packets until the last possible moment (i.e. when the network divides) and so duplicate packets can be avoided on all network links and segments [Deering and Cheriton, 1990].

Ideally, multicasting might reduce the total network bandwidth for MASSIVE-1 (in kbyte/sec) to

$$B = N(0.2M + 6.0) \qquad \text{(Equation 6-3)}$$

This is shown in figure 6.6 on page 90 which is directly comparable to figure 6.4 on page 86 except that it assumes multicast rather than unicast networking for inter-peer communication.

Fig.6.6: total network bandwidth against number of participants for a range of group sizes assuming ideal multicasting

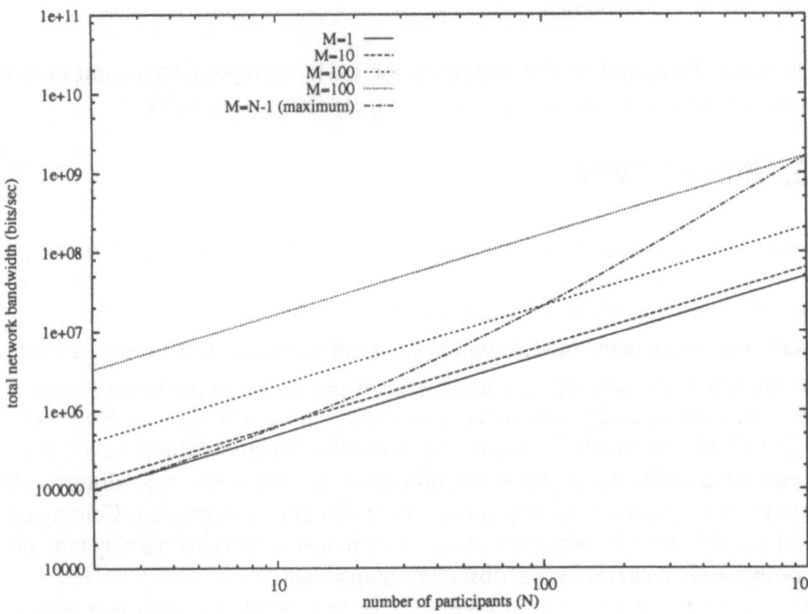

All updates (movement and audio) are assumed to be ideally multicast (as defined below). The remaining much reduced $O(NM)$ term corresponds to forming the initial association with a new peer and exchanging information such as appearance. There are a number of optimistic assumptions in this new model which would tend in reality to increase the $O(NM)$ term of equation 3. These are listed below.

- It ignores the possibility of individually tailored inter-peer communication which is inherently $O(NM)$ (this might include aspects of awareness negotiation for example or subjective views as discussed in [Snowdon et al., 1995]).
- It ignores the multicast management overhead associated with joining or leaving groups.
- It assumes reliable multicasting for at least some updates (though probably not audio) with no overhead or per receiver growth in traffic.
- It depends strongly on the rate at which new peer associations are formed, which will be application and context dependent.

This analysis also assumes (as does the model already presented) that total network traffic is the most important factor. This assumption is more realistic in the unicast case. Its limitations in the multicast case will be dealt with (among other issues) in the second part of this thesis.

Nevertheless, the potential reduction in total network bandwidth is dramatic: for an average group size of 10 it gives a six-fold reduction in required bandwidth, or equivalently six times as many participants supported by the same bandwidth. For larger group sizes the benefits are greater still, up to a factor of more than 20 times. So, with respect to total network bandwidth the scalability of an objective CVE will typically be much lower if it is based on unicast networking than if it is based on multicasting. This is a key observation and motivation for the work presented in the second part of this thesis.

This completes the analysis of the potential scalability of unicast-based peer-to-peer CVEs and of CVEs based on spatial trading in particular. The final section of this chapter summarises the key results and conclusions and leads into part II of this thesis.

6.4. Summary and Conclusions

This chapter has described spatial trading and the implementation of the aura relationship in MASSIVE-1. The previous section evaluated this from the perspective of the network by developing a model of expected network traffic which combines the behaviour of the application with expected user behaviour based on trials held using the system. This section summarises the conclusions of this chapter with regard to spatial trading and unicast peer-to-peer networked CVEs. Finally, it draws together key elements of the first part of this thesis and leads into part II.

6.4.1. Spatial Trading

This chapter has described how the aura component of the spatial model can be understood and realised as a brokering or trading exercise in which objects with compatible and overlapping auras are introduced to one another. This has been formalised as the concept of "spatial trading", which combines aspects of attribute-based naming services as in ODP with the virtual reality technique of collision detection. The key distinctives of spatial trading compared to normal attribute-based trading are: the offer and request both persist for as long as the object exists; the criteria for matching offers and requests include a notion of distance or proximity; a request always expects to learn of all matching offers; and offers and requests may change many times over their lifetimes but typically each change will be small and incremental. The formulation of spatial trading is one of the main outcomes of this thesis.

The network traffic model developed in section 6.3 shows that the use of spatial trading reduces the total network bandwidth requirement from $O(N^2)$ to $O(NM)$ (where N is the number of simultaneous users and M is the average group size). For CVEs the spatial trading approach has significant advantages over just splitting participants into limited sized worlds in that: it allows gradual, natural and visible transitions between groups within the same world; it allows different media to be treated in different ways; it avoids the need for invasive system intervention such as barring access to busy worlds; and it supports flexible control and graceful degradation through interactive or automatic modification of aura.

Spatial trading is a powerful notion which can be applied not only to CVEs but also to other CSCW and distributed systems. For example, Rodden's [Rodden, 1996] application of the spatial model to broader classes of collaborative systems could use spatial trading in a similar way (though with network-structured rather than Cartesian spaces).

6.4.2. Unicast-Based CVEs

The network traffic model developed in section 6.3 indicates that communication between peer processes is the dominant cause of network traffic for unicast-based peer-to-peer CVEs. This is true even with relatively small numbers of active peers. The total network bandwidth requirement is $O(NM)$ where N is the number of simultaneous users and M is the average number of other users in aura range. For the protocols used in MASSIVE-1 changing from unicast to multicast-based communication could result in a reduction in total bandwidth requirements of over 80% for $M = 10$ and more than 95% for large values of M. For an "ideal" protocol (if such a thing could be created) the reduction in required bandwidth might approach a factor of M, i.e. $O(N)$ total network bandwidth.

For systems with large values of M the attractions of multicasting are almost irresistible. However it must be borne in mind that this assumes strictly objective worlds (without tailored communication to individual peers). It also ignores multicast management and reliability costs. Finally, the potential reduction in bandwidth depends on how often peer relationships change and on how the initial knowledge of peers is obtained. This is expanded further in chapter 10.

6.4.3. Bridge

This section concludes part I of this thesis and sets the context for part II.

Part I has described the spatial model of interaction and the MASSIVE-1 system which implements it. Particular attention has been paid to the user aspects of awareness and interaction and to the network requirements of unicast-based peer-to-peer CVEs. Part II builds on this to present proposed extensions to the spatial model of interaction and the more recent MASSIVE-2 system which implements this extended model. The work presented in part II concentrates on a number of key issues which have been raised by the evaluations in part I. These key considerations are listed below.

- A computational model of awareness, such as the spatial model of interaction, provides unique opportunities for realising flexible patterns of interaction. This is expected to become increasingly significant in larger and more populated virtual worlds (see section 5.3.2) and is a critical system requirement.
- The process of awareness negotiation needs to take account of the context in which interaction occurs, for example a closed room compared to open terrain (see sections 2.2.1 and 5.4.3).
- Scalability with respect to network bandwidth demands appropriate use of network multicast facilities (see section 6.3.6).

The primary goal of the work presented in part II is to improve the scalability of both the spatial model and of CVE systems such as MASSIVE-1. It has been argued in this thesis that social and computational perspectives must be brought together if computer based systems are to approach the richness of everyday social existence. To this end the third party object extension to the spatial model presented in the next chapter provide socially motivated support for richer, context-sensitive interaction in a way that can be exploited to increase scalability. The MASSIVE-2 system demonstrates how this extended model can also be realised by a complementary multicast-based network architecture. The evaluations in part II parallel those in part I in considering awareness from a user perspective and system scalability from a networking and computational perspective.

This concludes chapter 6 and part I. The next chapter presents the proposed third party object extension to the spatial model of interaction.

7. Third Party Objects

This is the first chapter of part II of this thesis. Part I (chapters 3 through 6) has described the spatial model of interaction and the MASSIVE-1 system together with user and network oriented evaluations. These evaluations have raised many issues concerning the scope and realisation of the spatial model and the network scalability of CVE systems. Two issues have been singled out as being particularly significant for this thesis: network scalability; and the effects of context on interaction. These form the focus of part II.

This chapter proposes a fundamental extension to the spatial model of interaction which is called "third party objects". This provides socially motivated support for richer, context-sensitive interaction in CVEs. This is the principle theoretical contribution of this thesis. Chapter 8 introduces the MASSIVE-2 system which is based on and implements these extensions in addition to the original spatial model of interaction. Chapter 9 describes the implementation of awareness negotiation in MASSIVE-2 and presents a preliminary evaluation from a user-oriented perspective comparable to chapter 5 (except that this work is more recent than that presented in part I and experience with it is correspondingly more limited). Potential system scalability has been an essential consideration in shaping the extended model described here. Chapter 10 describes how the MASSIVE-2 system realises the extended model through a complementary multicast-based network architecture. This allows MASSIVE-2 to exploit the unique affordances for scalability which are inherent in the extended model (see section 7.3). The evaluation of chapter 10 analyses this and other possible network architectures and uses of multicasting in CVEs.

The next section presents the proposed third party object concept and describes how it fits into the existing spatial model of interaction. Section 7.2 describes potential applications of this extended model beyond those possible with the original model (in addition a number of prototype applications are described in chapter 9). As a prelude to chapter 10, section 7.3 highlights the affordances of the extended model which can be exploited to increase potential system scalability.

7.1. Theory

The original spatial model of interaction (described in chapter 3) reasons about potential awareness between objects in diadic relationships. Consider for example objects *A* and *B* in figure 7.1 (a) on page 96. The spatial model concepts of medium and aura determine whether this potential relationship is actually considered as being significant. *A* and *B* then use their respective foci and nimbi to negotiate mutual levels of awareness. These levels of awareness quantify the importance of each object to the other. Or equivalently, awareness may be regarded as a measure of the desired "quality of service" (QOS) to be given to communication between *A* and *B*. Recall that awareness may be different in each direction and may be different in each medium (e.g. graphics, audio, text). Consider now the introduction of a third object, *C*, as in figure 7.1 (b) on page 96. The original spatial model considers every diadic relationship individually and independently so that the relationship *A-B* is unaffected by the introduction of *C*. The extensions to the spatial model proposed in this thesis concern the role of "third party" objects such as *C* and the effect which they can have on other direct awareness relationships such as that between *A* and *B*. The concept of adapters in the original model can be seen as a limited and special case of third party objects and is considered in section 7.2.

Fig.7.1: diadic awareness relationships and third party objects

(a) a direct awareness relationship (b) a third party object, *C*

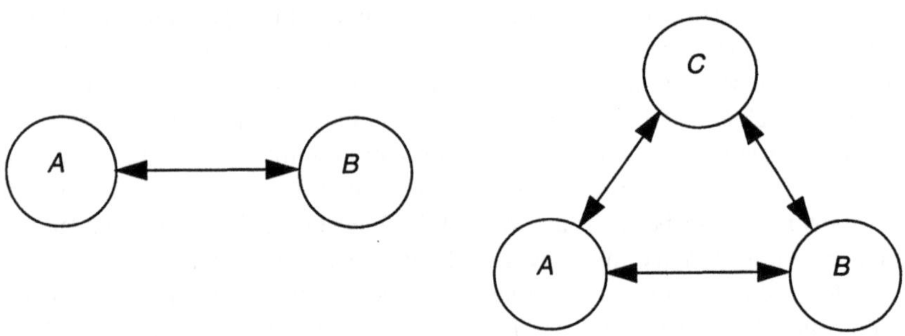

The third party object, *C*, may be used to represent some aspect of the context in which *A* and *B* interact, such as a room or a shared artefact. Involving the third party object in the awareness negotiation process allows for richer, context-sensitive patterns of interaction. This is one of the primary motivations behind this proposed extension to the spatial model of interaction (together with scalability). The remainder of section 7.1 describes third party objects in terms of their *effects* and their *activation*, i.e. *what* they do and *when* they do it.

7.1.1. Effects

Third party objects can have two basic effects on awareness which are termed "adaptation" and "secondary sourcing". These are described below and illustrated in figure 7.2 on page 97. These effects can be combined to create complex and flexible awareness relationships.

Fig.7.2: third party effects

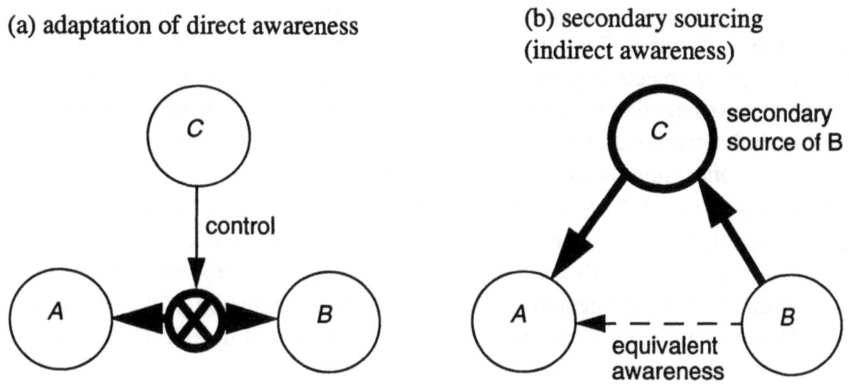

- Adaptation. Third party objects can modify or "adapt" existing awareness relationships. So in figure 7.2 (a) on page 97 *C* can affect *A* and *B*'s direct awareness of each other. It may increase awareness, leave it unchanged, attenuate awareness or eliminate it entirely. Different effects may be applied in each direction and in each medium.
- Secondary sourcing. Third party objects can introduce new *indirect* awareness relationships. So in figure 7.2 (b) on page 97 *C* can provide *A* with information about *B* and hence awareness of *B* even when *A* has no *direct* awareness of *B*. Secondary sourcing introduces new expressiveness into the model because *C* can act as a secondary source for a *group* of objects as well as for a single object. This allows groups of various kinds to be directly represented in the model. Typically a secondary source consumes information from a number of objects, filters and combines this information in some way and redistributes it. Different media may support different forms of secondary sourcing. For example audio signals may be mixed together, video images may be tiled and multiple geometries may be composited or approximated (see for example projection aggregation [Singhal and Cheriton, 1996] as a form of secondary sourcing and abstraction). The addition of secondary sources to the model also supports other forms of indirect awareness such as cross-medium adaptation (as seen in the text-to-speech convertor and the message board of MASSIVE-1, section 4.2) and other forms of delegation and representation.

The next section describes how these effects can be controlled. Section 7.2 then considers the combination of different effects and patterns of activation in a number of theoretical examples.

7.1.2. Activation

The activation of third party objects is determined by the existing direct awareness relationships, specifically those which involve the third party object itself. So in figure 7.1 (b) on page 96 the effect of *C* on the relationship *A-B* will depend on the relationships *A-C* and *B-C* and the associated awareness levels. Because the activation of third party objects depends on awareness it can exploit the power and flexibility of awareness negotiation. In particular this means that one third party object can control or influence another through the same third party mechanism applied to the first object's controlling awareness relationships.

Returning to figure 7.1 (b) on page 96 there are six potential awareness relationships between *A*, *B* and *C in each medium*: there is *A*'s awareness of *B*, *A*'s awareness of *C*, *B*'s awareness of *A*, *B*'s awareness of *C*, *C*'s awareness of *A* and *C*'s awareness of *B*. In principle *C* might be activated based on any combination of these awareness values. However three cases are particularly interesting and useful. These are illustrated in figure 7.3 on page 98 and are described below.

Fig.7.3: activation patterns for third party objects (*A*'s awareness of *B*)

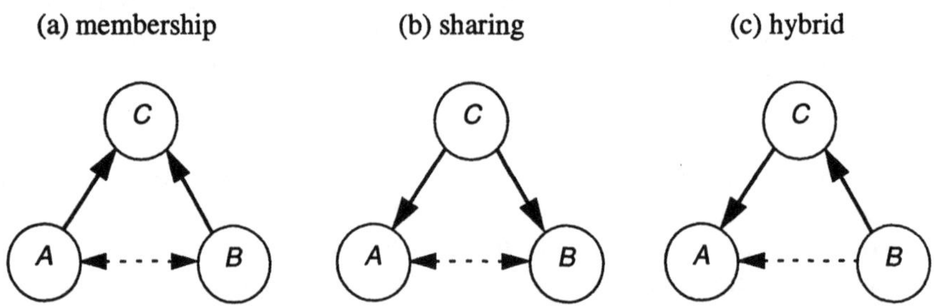

- Membership. This depends on the third party's awareness of both of the other objects (figure 7.3 (a) on page 98). In this case *C*'s awareness of *A* represents the degree of *A*'s *membership* of the group or set represented by *C*. Membership may be determined according to awareness in a normal medium (e.g. graphics or audio) may depend on a specialised medium (such as the adapter medium in MASSIVE-1, section 5.1.3), or may depend solely on the spatial medium (i.e. where things are). For example an object or a participant may become a "member" of a room by entering it. This might allow them to have full awareness of, and so interact fully with, the other occupants (members) of the room.
- Sharing. This depends on the awareness which both objects have of the third party object (figure 7.3 (b) on page 98). *A*'s awareness of *C* describes *A*'s interest in *C*. So if *A* and *B* are both highly aware of *C* then they can be regarded as shar-

ing *C*, in some sense, i.e. *C* is an object of common interest and may reflect a more general overlap of interests between *A* and *B*. Typically the awareness relationships between *A* and *B* would be enhanced in this situation to reflect the increased scope for cooperation or conflict represented by *C*. As in the case of membership, sharing might be assessed using normal media, using more specialised media (e.g. representing specific kinds of activity) or using space alone.

- Hybrid. This is a combination of the above and depends on one object's awareness of the third party and on the third party's awareness of the other object (figure 7.3 (c) on page 98). For example, *A*'s awareness of *B* would depend on the combination of *A*'s awareness of *C* and *C*'s awareness of *B*. This corresponds to *C* acting as a secondary source for *B*: conceptually information flows from *B* to *C* according to *C*'s awareness of *B*; it then flows from *C* to *A* according to *A*'s awareness of *C*. Flexible forms of abstraction and group level of detail can be realised based on this hybrid awareness. When *A*'s awareness of *C* is low *C* may provide an abstracted (indirect) view of *B* and any other members. When *A*'s awareness of *C* is high *A* may be directly aware of *B* and the other members of *C*.

7.1.3. General Observations

Third party objects, as well as having their own particular characteristics, are also first class objects within the extended spatial model of interaction. This has three important implications which are listed below.

- Third party objects can themselves exploit focus, nimbus and aura to manage their own interaction and operation.
- Third party objects can affect one another, allowing the construction of combined, linked and nested patterns of effects.
- Third party objects, like any object, can be dynamic. For example, in a CVE they might be dynamically introduced into the system, be mobile and change in size and effect over time.

This completes the definition of the third party object extension to the spatial model of interaction. The next section gives some examples of the applications and operation of third party objects which illustrate and motivate this extension.

7.2. Examples

Third party objects can be used to create many different forms of context-sensitive interaction in CVEs. A number of possible applications are described below: rooms and buildings, cells and zones, crowds, floor control, common foci, cross-medium adaptation, spatial model adapters, remote awareness and load-management. Chapter 9 includes details of a number of working examples implemented in MASSIVE-2.

Rooms and Buildings

A third party object could be used to create and represent a room or building in order to structure interaction in terms of the space in which it occurs. A room third party might determine its members according to spatial containment. It could then enhance the direct awareness between members to give high-fidelity audio, for example. It might also attenuate or entirely cut off awareness across its boundary in one or both directions so that participants within the room could be protected from interruption and/or eavesdropping. Simultaneously the room might act as a secondary source, providing an overview or summary of the activity within the room, e.g. the number and identity of its occupants.

Note that third party objects could be applied recursively, so a building might contain floors which might in turn be composed of rooms. Buildings, floors and rooms could all have their own unique effects on awareness. Similarly, other kinds of third party object (such as those described below) could be included in these spaces to further tailor the patterns of interaction.

Cells and Zones

Cells or zones are found in a number of multiuser VR systems (e.g. NPSNET [Macedonia et al., 1995]). These are used to scope interaction and awareness in open spaces, i.e. they correspond to the general properties of distance and interaction (as in "too far to see clearly", etc.) rather than to world content such as rooms and buildings. A cell can be expressed in terms of a simple third party object which provides (indirect) awareness of its content. This is directly comparable to the use of a multicast group as an abstraction for grouped indirect communication. In a practical system (such as MASSIVE-2) cells may not actually be represented in terms of third party objects, but at a more "primitive" level within the distribution framework.

Third party objects which are defined spatially (such as cells, rooms and boundaries) are normally referred to in this thesis as *regions*. The spatial limit of membership defines the region's *boundary* which is fundamental when considering a region's effects.

Crowds

A third party object may be used to represent a crowd or other distinguishable group of participants (see [Benford et al., 1997] for an extended discussion of crowds). A crowd third party object is more likely to be mobile and dynamic than the previous types, reflecting the nature of crowds. So a crowd third party object may follow its members about the virtual space, dynamically extending its effects to include them. The effects and activation of a crowd might be similar to a room, above, except that the crowd members are likely to have some common external object of interest. Consequently the crowd object might tend to enhance rather than attenuate its members' awareness of external events. However the crowd may still restrict an external observer's awareness of its members, for example it may provide a secondary sourced overview of its members as a whole (they may "merge into the crowd" from a distance). So a crowd third party could be used to reduce effective world complexity as seen by non-members, by replacing awareness of many individual crowd members with awareness of a single crowd object. A crowd might also provide tailored forms of interaction for its members, reflecting their common identification and interests, for example alternative styles of navigation and enhanced awareness of objects of common interest.

Floor Control

A third party object could realise traditional floor control mechanisms found in CSCW (e.g. [Sarin and Greif, 1985]) by cutting off direct awareness between participants and providing a single secondary sourced view of the interaction which conforms to the appropriate floor control policy (for example round robin, loudest wins or chosen by chairman).

Common Foci

A third party object activated by sharing could enhance awareness between participants who are focusing on it. This enhanced awareness would reflect the increased scope for cooperation and conflict represented by the common object of interest. This might be important in the context of shared design or shared information visualisation. This reflects the kind of spontaneous interactions which can occur in the everyday world because of sharing physical artefacts (which leads to co-location and hence to the opportunity for communication).

Cross-Medium Adaptation

Third party objects can be used to realise cross-medium adapters such as MASSIVE-1's message board and text-to-speech convertor (sections 4.2.1 and 4.2.2). These consume information from one set of objects in one medium, transform it, and make it available in another medium, i.e. they are acting as secondary sources. Such objects can be created within the confines of the original spatial model of interaction however, realising them explicitly as third party objects has two important advantages: secondary sourcing can be combined with adaptation to reconcile direct vs. indirect awareness; and it provides a framework to allow secure and controlled delegation and distribution rather than an ad hoc arrangement visible only to the user.

Spatial Model Adapters

An adapter in the original spatial model modifies interaction by changing one object's foci, nimbi and auras. The activation of adapters was not specified in the original model but might be based on the object's awareness of the adapter (as in MASSIVE-1). In this case an adapter can be viewed as a degenerate form of third party which provides adaptation but not secondary sourcing and which is activated according to one object's awareness of the adapter (the other object is not involved) and so the effects of the adapter are the same as viewed by every other object.

Remote Awareness

A third party object could be used in its secondary sourcing mode to provide indirect awareness at a distance, i.e. it might consume information in one place and redistribute it in another location, possibly even in a different virtual world. This could also be applied to portals to create "dynamic portals" (which give awareness of their destination) as in BrickNet [Singh et al., 1994]. As with cross-medium adapters this could be realised to a limited extent in the original model but with no basis for management, control or accountability.

Load-Management

The final application of third party objects described here is in managing system load (and possibly also network load). Third party objects such as rooms, buildings, cells and crowds create additional structure within the virtual world and can scope and constrain awareness. So, in the face of increasing machine or network load, a system could dynamically introduce or reconfigure third party objects. Because they localise and limit interaction these third party objects can help to control system load (this is explored more directly in the next section). For example the system might perform dynamic clustering of participants and world content and introduce corresponding third party objects, partitioning the environment into more manageable units. With the inclusion of secondary sourcing and abstraction, partial awareness of the contents of other clusters and groups would be preserved. Third party objects may also have a role in load balancing, for example identifying objects which are "related" or likely to interact and which should be brought together on a single machine.

The last two sections have described the concept of third party objects and provided a number of illustrative examples. The next and final section of this chapter describes how this extended model can be exploited to increase potential system scalability.

7.3. Exploitability

This third party object extension allows much richer representation of the influence of context on interaction than did the original spatial model of interaction. This support for interaction context is the socially motivated side of the third party object model. System scalability (in terms of computation and communication) has also been a key motivation in forming the extended model described here. This aspect of the spatial model with third party objects is the subject of this final section of chapter 7. The exploitation of third party objects to enhance scalability in MASSIVE-2 is described in chapter 10.

Increased scalability is possible with third party objects because adaptation can be used to suppress potential awareness relationships in a way which is consonant with the nature and structure of the collaborative environment. For example, a room or building with opaque (sound-proof, etc.) boundaries causes awareness relationships which cross the boundary to be suppressed with a corresponding potential reduction in requirements for communication and computation. This allows interaction to be localised within closed spaces (rooms, buildings) and also allows closed spaces to be ignored by external participants.

This is useful in itself, but its applicability is greatly extended by the inclusion of secondary sourcing. For example, the third party object may simultaneously suppress direct awareness and provide an abstracted (less costly) overview of the suppressed activity. A crowd (see previous section) is a good example of this. Secondary sourcing also allows other kinds of third parties such as rooms and buildings to support a limited degree of awareness. Finally, it allows the creation of cells or zones which provide reduced information from a distance but full information when nearby (for example, a third party may use hybrid activation to switch between adaptation and secondary sourcing according to awareness of the third party object).

So, in terms of potential scalability there are three situations in the extended spatial model of interaction including third party objects in which one object cannot possibly be aware of another object (and therefore communication and computation might be avoided). These three situations are:

- when the objects are outside of aura range, as in the original model;
- when the objects are on opposite "sides" (i.e. inside and outside) of an opaque boundary such as the edge of a closed room or building; and
- when one object is within a hybrid activated third party such as a crowd and the other object is outside and sufficiently distant (i.e. with sufficiently low awareness).

The first case addresses scoping of interaction in open spaces. The second addresses scoping of interaction based on significant world structure and content. The third affords additional scalability through the introduction of reduced-detail group abstractions. All three are exploited in the MASSIVE-2 prototype (see chapter 10).

7.4. Summary

This chapter has presented the third party object concept as an extension to the original spatial model of interaction (as reviewed in chapter 3). Third party objects can affect awareness and interaction through a combination of *adapting* existing direct awareness relationships and introducing new indirect awareness relationships through *secondary sourcing.* A third party object is controlled through the awareness relationships which exist between the third party and the other objects in the space. In particular activation of third party objects may reflect *membership, sharing* or *hybrid* awareness relationships. Third party objects have many potential applications including structuring interaction through rooms, buildings and cells, support for crowds, common foci, floor control and load management. In addition, third party objects can enhance system scalability by scoping interaction based on significant world structure and content and through the introduction of group abstractions.

This model has been implemented in the MASSIVE-2 CVE system which is introduced in the next chapter. Chapter 9 describes the implementation of awareness negotiation for the extended spatial model in MASSIVE-2 and presents a number of example applications which demonstrate the model's potential. Chapter 9 also presents a preliminary evaluation of third party objects in use. Finally, chapter 10 describes the multicast communication architecture of MASSIVE-2 and the way in which it exploits the extended spatial model as discussed in the previous section. The evaluation in that chapter focuses on system scalability and the use of multicasting in CVEs.

8.
MASSIVE-2

This chapter describes the design and implementation of MASSIVE-2, the second of the CVE system prototypes presented in this thesis. The primary motivations behind this system were to:

- retain and build on the basic concepts of the spatial model as demonstrated in MASSIVE-1 (described in chapters 3 through 6 of this thesis);
- prototype and experiment with the third party object extension to the spatial model proposed in the previous chapter; and to
- realise this through appropriate use of multicast network communication for enhanced network scalability (motivated by section 6.3).

MASSIVE-2 goes significantly beyond MASSIVE-1 in breadth and generality and includes many features which were omitted from that first prototype as well as completely new elements. Both systems support interaction via text, 2D and 3D graphics and packetised audio and use awareness to control interaction in all media. In addition, features which have been introduced in MASSIVE-2 include:

- implementation of the proposed spatial model extension of third party objects, including adaptation and secondary sourcing;
- use of network multicast communication for update messages such as movement and audio;
- an effective mapping between spatial model extensions and multicast groups to provide dynamically controlled, structured and appropriate communication;
- a well-defined API for creating applications; and
- a richer framework for object behaviour including direct manipulation of virtual objects.

This chapter provides a general introduction to MASSIVE-2, comparable to chapter 4's introduction to MASSIVE-1. Details of the most important aspects of the implementation are deferred until chapters 9 and 10. The first two sections of this chapter are primarily user-oriented. Section 8.1 describes the interfaces which are presented to a normal system user, i.e. a participant in the virtual environment. Section 8.2 then describes the kinds of third party objects which are available. The last two sections deal with the implementation of MASSIVE-2. Section 8.3 describes the distributed programming model adopted which differs significantly from that adopted for MASSIVE-1. Finally, section 8.4 gives an overview of the software and network architecture of the system. This architectural overview is an introduction to the implementation descriptions in the next two chapters.

At the outset the author wishes to acknowledge, but also delimit, the contributions of various colleagues in the Communications Research Group at the University of Nottingham to the development of MASSIVE-2 (which is also known as CVE, the Communications research group Virtual Environment system). Dr David Snowdon, author of the AVIARY system [Snowdon and West, 1994], has been the closest collaborator. In particular he has been responsible for the platform independent 3D graphics library, low-level device interfacing (e.g. to trackers) and many of the basic library elements such as lists, hash tables and error handling. Michael Fraser has provided cosmetic enhancements to the graphical user client (and is continuing to use and extend aspects of the user interface as part of his program of study for the degree of PhD); he has been supported in this by Jolanda Tromp. Dr Snowdon and Dr Marcus Roberts contributed to the design of the distributed object system used to implement MASSIVE-2, and Dr Roberts provided the ODP-style trading facility. Thank you.

8.1. User Interface

This section describes the way in which MASSIVE-2 is presented to a normal user. The appearance of the system is described and also the types of control and interaction which are made available to the user.

MASSIVE-2 does not support the same degree of heterogeneity of user machine capabilities as did MASSIVE-1. The minimum requirement is a 2D bitmap graphics terminal with keyboard and mouse (e.g. an X-terminal). It is anticipated that the system will normally be used with 3D graphics and this is reflected in the design of the combined 2D/3D graphical user interface which is shown in figure 8.1 on page 107. At the left of this main window is the 3D graphical view (this is blank when 3D graphics is not available on the client machine). This shows a view into the virtual world equivalent to the graphical client of MASSIVE-1. At the right of the window is a simple line-drawn map which replaces the text character map of MASSIVE-1's text client. The map is always centred on the user and can be zoomed in and out. Three small indicators towards the right of the window give feedback on current audio status to assist the user in making themselves heard. They indicate that audio is being sent, that the microphone volume is adequate and that the volume is excessive. The remainder of the interface makes available the various control options which are described below.

A user can navigate in three ways: using the cursor keys, using the buttons at the bottom left of the interface and using the "mouse vehicle" in the 3D graphical view.

Fig.8.1: MASSIVE-2 graphical user interface

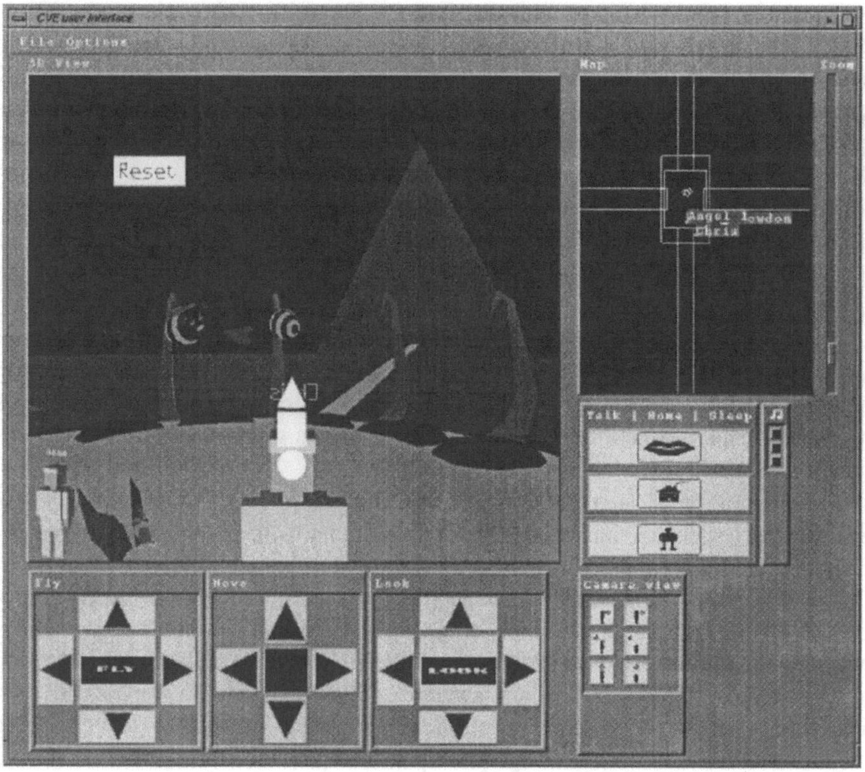

The mouse vehicle is based on the normal navigation facilities provided by the DIVE distributed virtual environment system from the Swedish Institute of Computer Science (see [Carlsson and Hagsand, 1993] and [Hagsand, 1996]) and comprises a triangle above a square above a circle. Dragging each of these shapes using the mouse allows navigation in two of the six available degrees of freedom. For example clicking the mouse on the triangle and dragging up screen causes the user to move forwards and dragging it to the right causes the user to turn to the right. In addition the "reset" button in the 3D view and equivalent buttons in the 2D interface reset the user's orientation to horizontal (on the ground plane).

MASSIVE-2 provides two other less direct means of moving about a virtual world: conveyors and a "home" facility. Conveyors are defined objects or regions within a virtual environment which cause the user to drift in a particular direction (the user can easily override this drift using the other navigational facilities). Conveyors can be provided to assist novice users and to encourage participants to move towards key elements within a virtual world. In addition each user has a "home" position within the virtual world. At any time they can press the home button in the interface to move to this home position. If the home button is pressed for long enough then they return towards their home position and orientation. This allows a user who becomes lost or disorientated to return (with continuity) to a familiar and stable reference point.

At present MASSIVE-2 provides less support for gestures than did MASSIVE-1. The "sleep" gesture has been retained as essential for managing the relationship between activity in the real and virtual worlds (see section 5.3). A graphical "mouth" has also been retained to indicate that the user is speaking. This is represented by a speech balloon which appears above the user's head. As in MASSIVE-1 this helps to identify speakers and to diagnose audio and network problems.

Audio in MASSIVE-2 is presented to the user in stereo, with the left-right pan determined by the location of the audio source with respect to the user. In addition, as in MASSIVE-1, the volume is determined by the user's level of awareness of the source.

There is a text window which allows text messages to be exchanged but this is not normally used. It is expected that all users will have full audio-graphical interaction facilities, though on machines with a range of processing capabilities. This reflects the much greater availability of audio and 3D graphics support on current workstations and PCs.

Presentation of 2D and 3D graphics, text and audio is all based on awareness level which is calculated independently for each medium. 3D graphics can be dynamically modified according to awareness level to give awareness-driven level of detail (this only applies to single objects and should not be confused with secondary sourcing and abstraction which can apply to groups of objects). As noted above, audio awareness level controls audio playback volume. Text and 2D graphics are presented or not depending on the text medium awareness level (text and 2D graphics are considered to be aspects of the same medium with regard to awareness - text messages provide communication while 2D graphics provide spatial information).

MASSIVE-2 supports direct manipulation of virtual objects. Using the mouse a user can click on an object in the 2D or 3D graphical views and attempt to drag it. The request is communicated to the object involved which makes its own response. For example, it may choose to relocate itself according to the user's request so that the user effectively moves the object about. Equally, it may make no response at all. Alternatively it may change its behaviour or trigger some aspect of an application. For example, the mouse vehicle comprises three local graphical objects which respond to manipulation by translating and rotating the user's own embodiment.

8.2. Third Party Objects

The last section described the normal user view of MASSIVE-2. This section describes the kinds of third party objects which can be realised in the system. Section 9.2 in the next chapter describes some applications and demonstrations which make use of these third party objects.

In MASSIVE-2 any object within the virtual environment (i.e. any *artefact*) can be a third party object. Referring to section 7.1 these third party objects can have both adaptation and secondary sourcing effects and can be activated based on membership or on hybrid awareness. In both cases of activation the third party's awareness of another object is based solely on the spatial medium, specifically whether the object is fully contained within the third party. However an object's awareness of the third party (used in the hybrid case) is based on whichever medium is being considered. A number of the available third party objects combine both forms of activation. At present activation through sharing is not supported.

MASSIVE-2 allows completely customised third party objects to be created within the constraints of the implementation (described in section 9.1). In addition a number of standard types of third party object are available to world designers and application builders. These are: closed room; open region; level-of-detail region; panopticon cell; and crowd. These standard third party objects are described below.

- Closed room. The closed room third party can be used to implement rooms, buildings and other closed structures. The effect of a room third party is to suppress awareness between members and non-members both in and out. So participants within the room can interact with each other normally but they are not aware of events outside the room. Conversely participants outside the room are not aware of people or events within the room. This is illustrated diagrammatically in figure 8.2 (a) on page 110. Note that only awareness relationships from top to bottom are shown in this figure. The third party object's appearance is the room itself. Room-type third parties can be nested to create hierarchically structured spaces such as rooms within floors within buildings.
- Open region. An open region third party object has no effect on awareness. It is used to create area of interest cells such as those found in NPSNET [Macedonia et al., 1995]. The operation of open regions is described in chapter 10 which describes the distribution and replication scheme used in MASSIVE-2. An open region is shown in figure 8.2 (b) on page 110.
- Level-of-detail region. The effect of a level of detail region is based on hybrid or indirect awareness. When an observer has a low awareness of a level of detail region then their awareness of its members is suppressed by adaptation. However when an observer is sufficiently aware of the level of detail region then they have normal awareness of its members while awareness of the region itself is suppressed. So a level of detail region provides an indication of the *presence* of a region at low levels of awareness, but "opens up" to reveal its *contents* at higher levels of awareness. This is shown in figure 8.2 (c) and (d) on page 110 for low awareness and high awareness respectively. Note that, unlike a closed room (but like a crowd, below), a level of detail region has no effect on the awareness which its members have of objects outside the region, i.e. they can see (hear, etc.) objects outside the region.
- Panopticon cell. This is a version of Jeremy Bentham's panopticon cell as described in [Foucault, 1977]. This is an asymmetric closed region in which members have no awareness of external objects but which allows external observ-

Fig.8.2: standard third party object types and effects in MASSIVE-2 (only awareness from top to bottom is shown)

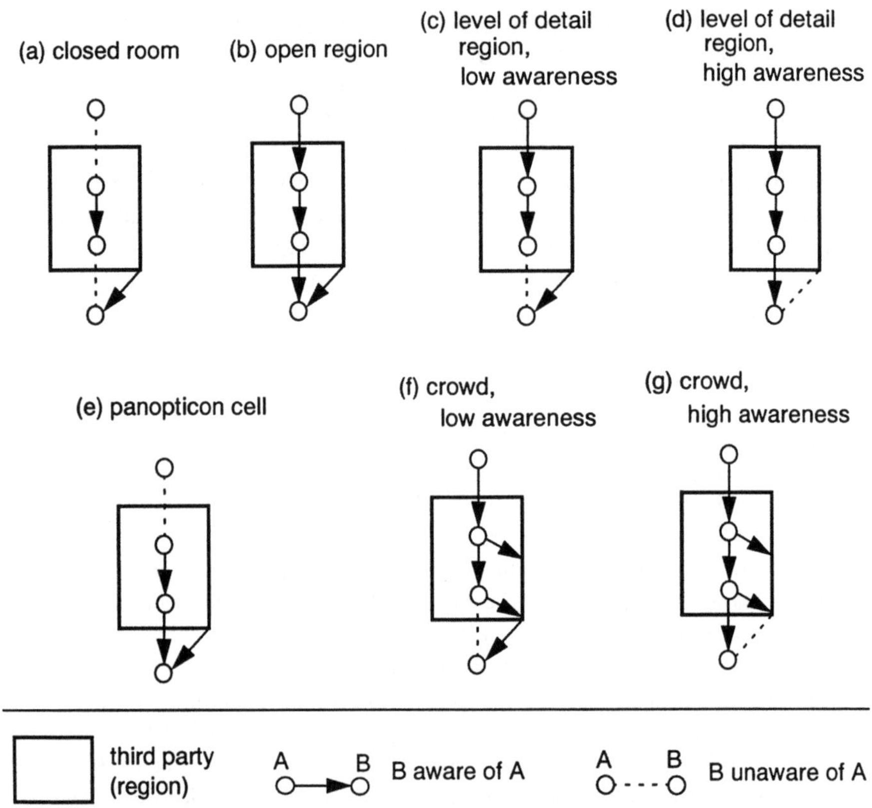

ers full awareness of objects (including participants) within the cell (see figure 8.2 (e) on page 110). Bentham proposed that a prison built on these principles would cause the inmates to learn to control their own behaviour because of the threat of constant observation. He also believed that it might provide a tool to establish a new science of the study of human behaviour. A more mundane application is the unobserved monitoring of users in a general virtual environment.

- Crowd. The crowd is an extension of the level of detail region to include dynamic secondary sourcing (the level of detail region, above, has a fixed appearance). When an external observer is sufficiently aware of the crowd third party in a given medium then they have direct awareness of the crowd's members in the normal way while their awareness of the crowd itself is suppressed. However when the observer's awareness of the crowd is low then they are aware of the crowd but not of its individual members. The crowd changes its appearance to represent information about its current members. The (simple) default crowd changes the size of its 3D graphical appearance in proportion to the number of

members. Its 2D map representation shows a count of the current number of members. Its audio representation is a real-time mix of the audio from all of its members with additional optional filtering. For example, a low pass filter can be used to muffle the crowd sound to indicate to observers that they are hearing a crowd. A crowd third party is illustrated in figure 8.2 (f) and (g) on page 110 for low awareness and high awareness, respectively. Note that if presentation of the crowd were not suppressed when an observer was directly aware of its members then they would hear the sounds made by the individual members followed a short time later by the crowd's secondary sourced "echo". This demonstrates the need for explicit management of secondary sourcing.
- Closed crowd. A variant of the crowd is also available which always provides the abstracted view to non-members irrespective of their awareness of the region. This corresponds to the case in figure 8.2 (f) on page 110 but does not depend on awareness.

Some of these standard third party object types are visible in the demonstrations in section 9.2.

This section and the previous one have described MASSIVE-2 from the perspective of a normal user. The next two sections provide background and introduction to the implementation details included in the next two chapters. The next section describes the distributed object model used to implement MASSIVE-2. Finally, section 8.4 gives an overview of the system's network and software architecture.

8.3. Distribution Model

This section describes the model of distributed computation adopted for this second phase of prototyping; in the next section this model is used to give an overview of the design of MASSIVE-2.

The distribution model for MASSIVE-1 was a variant of the connected component model of distributed programming, and was described in section 4.3. However for MASSIVE-2 an alternative style was adopted based on distributed objects. The main motivations for this change of style were that:

- the connected component model of MASSIVE-1 was suited to a realisation based on unicast more than multicast network communication; and
- the lack of explicit support for objects in MASSIVE-1 made programs hard to structure, understand and extend (the previous style was based on "floating" interfaces).

There were also other shortcomings of the first system's implementation (rather than the concepts as such), for example the lack of support for multiple threads of execution.

Some of the key characteristics of the distributed object model used for MASSIVE-2 are outlined below. This is not a complete description but is sufficient for the implementation discussions which follow.

- The object system is class-based (rather than prototype-based).
- Each object exists within the context of a single operating system process (e.g. a UNIX process). Many objects may exist within the same process (in which case they are described as being local to one another).
- All objects (whether local or remote) are identified by a 20 byte globally unique object identifier. Program semantics for correct operations are identical for both local and remote objects (although remote actions are more likely to fail).
- Message despatch is based on an 8 byte hash value derived from the method name and argument names and types (more like Smalltalk or Objective-C than C++). This gives a general basis for polymorphism and avoids the need for global message type registration.
- Local objects can be specified to act as proxies for remote objects: messages sent to the remote object will be transparently despatched to the local proxy if it has an appropriate method, otherwise a remote message send will be made.
- Half of the object name space is given over to multicast groups. Messages can be sent to multicast groups in the same way as to individual objects.
- Any number of objects can independently join a multicast group as receiving members. A message sent to the group will then be despatched to all of its receiving members (local and remote) as if the message were sent to them directly.
- An object can send a message to a multicast group whether it is a receiving member of that group or not.
- Each message send to an object (but not to a group) can be either an asynchronous message send or a synchronous (remote) procedure call. All message sends to groups must be asynchronous.
- Asynchronous message sends can be individually specified to be reliable or unreliable and to be source-ordered or unordered; these four classes of service are directly supported by the underlying communications (both unicast and multicast).
- Multicast groups are supported via a minimal protocol running over IP multicast [Deering, 1989], which provides optional source ordering and reliable delivery on a per-message basis (reliable delivery is based on message sequence numbers, finite heart-beat messages with exponential timer back-off and receiver driven retransmission).

As was the case with MASSIVE-1 the distribution system and communications library were written specifically for this prototype system. As can be seen from the list above much of the system is generic. However some of the features described above are less standard but are important for the realisation of MASSIVE-2. These are: support for local proxies of remote objects (which also relies on polymorphism);

integration of multicasting; and support for different classes of service for remote message sends.

The next section introduces the implementation of MASSIVE-2.

8.4. Implementation Overview

This section introduces the implementation of MASSIVE-2 and gives an overview of the system's network and software architecture. This is further developed in chapters 9 and 10 which give additional details of the system's implementation of third party effects and its management of multicast communication, respectively. This section begins by describing the processes which cooperate in a complete MASSIVE-2 session. It goes on to describe the four main distributed object classes on which the system is based.

8.4.1. Processes

A MASSIVE-2 multi-user session involves a number of simultaneous participants sharing a common virtual world. Four types of application are involved is such a session and each one executes as an operating system heavyweight process. These applications are listed below.

- A *node manager* runs on each computer involved in the session. This is a standard service of the distributed object system and is not specific to MASSIVE-2. The node manager provides an ODP-style [ITU-T, 1995] trading service, plus message forwarding and redirection services to support object migration.
- One or more *world servers* maintain the world and its normal content. A world server normally registers its existence with a node manager. This allows users to locate worlds without needing to know native object identifiers. Many worlds may exist concurrently. Only one process *creates* each world but many other processes can add artefacts to the world and interact in it.
- One *graphical client* and normally one *audio client* is required for each participant. The use and presentation of these was described in section 8.1.

These applications are not "special" in themselves: they are hosts for particular objects created using the distributed object system. For example, the node manager application hosts a *PlodNodeMgr* object which provides methods for performing attribute-based trading. The node manager process is the only process which requires a well-known and globally consistent UDP port for communication. The next section describes the fundamental object classes used in MASSIVE-2 and describes how they communicate and interact.

8.4.2. Classes

MASSIVE-2 is implemented using four core object classes: *World, Artefact, Aura* and *Group*. Each *World* object represents a single virtual world and is the initial point of contact on joining that virtual world. *Artefact* objects represent artefacts within the virtual world such as rooms, parts of a visualisation or the embodiments of users or agents. *Aura* objects represent spatial model auras and are used for spatial trading. *Group* objects manage artefact replication between cooperating processes and coordinate the use of multicast communication. Also, as described in the following chapters, third party objects are realised by *Group* objects in coordination with *Artefact* objects. Figure 8.3 on page 114 shows an overview of the processes and objects comprising a minimal session, with one node manager, one world server process (containing one master *Artefact*) and a single user client process (with a single embodiment *Artefact*). *Artefacts* are replicated on demand (as described below and in chapter 10). In the figure *Artefact A* in the world process has been replicated in the user's client process while the user's embodiment, *Artefact B*, has been replicated in the world server process (though probably only in the spatial medium). Distribution of messages to replicas is via multicast groups (as described at the end of this chapter). The classes and their interaction are described in more detail in the following sections.

Fig.8.3: overview of MASSIVE-2 network and software architecture

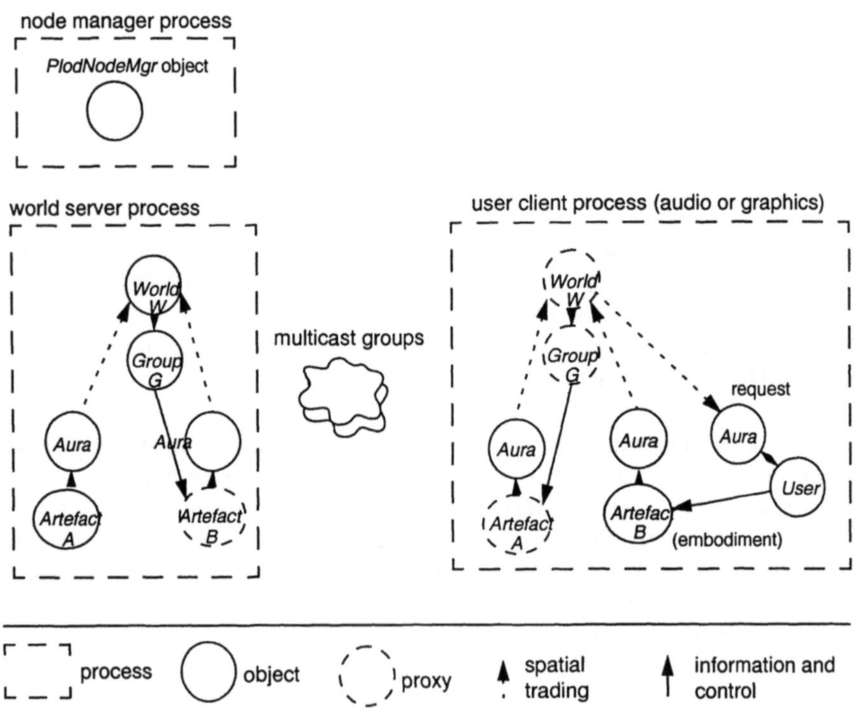

World

The concept of a virtual world is a fundamental one for CVEs. In the context of the spatial model of interaction they are disjoint spaces within which communication and interaction may occur. In MASSIVE-2 worlds are represented within the system by instances of the class *CveWorld*. This simple class acts as a repository for world-specific configuration information and as a single point of access at which version checking and access control can take place. The world-specific information comprises:

- identities of default communication channels (i.e. *Group* objects and multicast groups) associated with the world;
- identities of media supported by the world - different worlds may support different sets of media including text, 2D and 3D graphics, audio and video; and
- protocol version numbers for supported media - to ensure that interaction can occur safely within each world without using explicit type information.

The process of gaining access to a new world is illustrated in Figure 8.4 on page 115.

Fig.8.4: a user process accessing a world

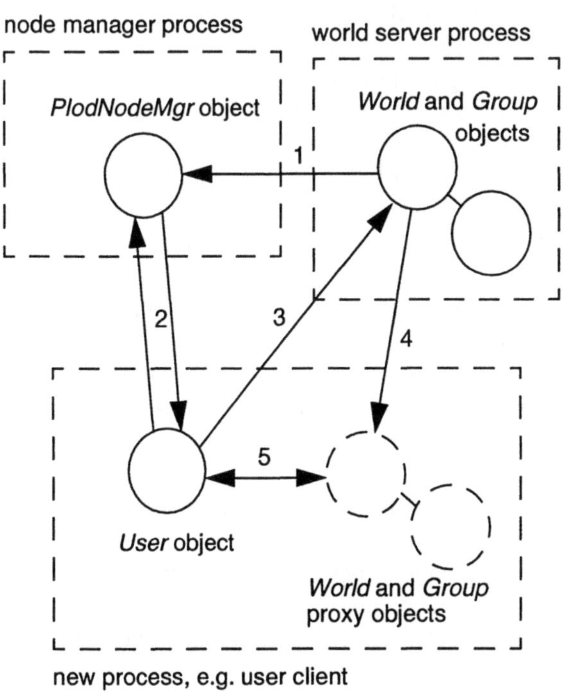

1. *World* object registers with node manager.
2. Look up *World* object identity with node manager.
3. Register with *World* master object, checking protocol versions.
4. Create local *World* proxy which replicates necessary world-related information.
5. Interaction with local *World* proxy (e.g. via *Artefacts* and *Auras*).

First, a world server process creates a new *World* object which is the entry point for that world. Normally this object's identity will be registered as an offer with the local node manager/trader (1). Then, when another process (or rather an object within another process) wishes to join this world it will look up the *World* object's identity (and network location) in the trader (2) and then register with that *World* object (3). Registration includes checking protocol version numbers and results in the creation of a local *proxy* for the *World* object in the new process (4) which has a copy of all of the world-specific configuration information and which handles subsequent access to the world in that process (5). This proxy makes use of the distributed object system's support for local proxies and run-time delegation to transparently act on behalf of the main (master) *World* object within the local process, reducing network traffic, latency and subsequent load on the master *World* object.

Note that there is a *Group* object associated with the world which handles key aspects of communication and *Artefact* replication for the world as described in the following sections. When the *World* proxy is created it also creates a local proxy for this *Group* object.

Artefact

As well as virtual worlds the other fundamental concept shared by all CVEs is that of artefacts (as they are referred to here) which may represent users, agents, applications or other objects within the virtual world. In MASSIVE-2 each artefact is represented by an instance of class *CveArtefact*. This is composed in turn from a number of medium-specific sub-components (using multiple inheritance in the current system). The standard media which may be supported by any artefact are currently:

- *spatial*, which includes information about position, orientation, size and solidity and is independent of appearance in any other medium;
- *text*, which (as in MASSIVE-1) allows an artefact to be named and to exchange text messages;
- *2D graphics*, which gives simple 2D appearances for artefacts, which is presented to participants in the 2D map view;
- *3D graphics*, which gives 3D geometries for artefacts as is typical in virtual reality systems (and which is significantly more flexible than in MASSIVE-1, being comparable to VRML-1 [Bell et al., 1995] with additional support for incremental modification of geometries); and
- *audio*, which allows real-time conversation through packetised audio.

An artefact's representation in a medium comprises a number of well defined attributes and in some cases one or more streams of medium-specific messages or events (e.g. text messages or audio packets). The *Artefact* class provides methods to set and query attribute values, to generate events and to register other object methods to be called when attributes change or events occur (i.e. "callbacks").

The basic model of interaction and communication in MASSIVE-2 is that an object or process (e.g. a user client) creates and manipulates *Artefacts* within the local process and associates these with a given virtual world. In this way it makes information about itself available. The *Group* objects cause these *Artefacts* to be replicated on demand in other processes which have joined the same virtual world. Other objects or processes can observe these artefacts and "reply" by manipulating other artefacts. So a typical process or application both manipulates artefacts within the virtual world (in order to communicate and interact) and monitors artefacts (e.g. displaying representations of them to a human user). In MASSIVE-2 it is normally the case that each artefact is (and remains) under the control of a single object or application or is under its own control (all artefacts are considered to be at least potentially active and autonomous). This may be contrasted with a more database-oriented view in which artefacts are regarded as passive blocks of data which may be arbitrarily modified and moved between processes and owners.

It is interesting to reflect that this may be one area in which MASSIVE-2 retains some of MASSIVE-1's particular emphasis on tele-conferencing: there are both social and technical reasons why artefacts representing users should be considered to be active components under the control of a single process. For example, there is a clear notion of ownership and identity between a user and their embodiment and there will also normally be links to specific hardware interface components, i.e. the machine which the user is currently using and any specialised hardware which comprises the interface (such as a head-mounted display and tracking devices).

When an artefact is placed into a specific virtual world the *Artefact* object "signs on" with the corresponding local *World* proxy object (creating the proxy if necessary). All monitoring and observation of artefacts is managed using local spatial trading. This depends on the *Aura* class which is described in the next section. As already noted, replication and distribution of artefacts to other processes (e.g. to other user clients) is handled by the *Group* class which is described last.

Aura

As in MASSIVE-1 a process or object learns which artefacts are present in the world by using spatial trading (see chapter 6). In MASSIVE-2 this is based on the *CveAura* class which represents a spatial model aura. The information which defines an *Aura* object comprises: world identity; the set of media to which it applies; position in 3D space; size as an axis-aligned cuboid extent; the identity of the artefact with which the aura is associated; and the aura's type (defined below).

Unlike the original spatial model MASSIVE-2 distinguishes between auras which correspond to offers of information and auras which correspond to requests for information. This distinction was included to a limited extent in MASSIVE-1 (the *type* attribute in table 6.1 on page 79) but is not explicit in the spatial model of interaction. This distinction is particularly important when dealing with large numbers of passive non-user objects, each of which has its own aura. In this situation it is unnecessary and wasteful to check for aura collisions between all of these passive object auras since none of them wishes to respond to the others. In many applications there will be many more passive artefacts than active ones.

In MASSIVE-2 every *Artefact* has an associated *Aura* object which represents its offer aura. This corresponds to a declaration that the artefact is present in the world. These *Aura* objects are created and managed by the local *World* proxy object. This maintains a spatial octree for the world in which it keeps track of all offer auras. In addition, any artefact or other object (such as a user object) which wishes to find out about other artefacts in the world will create one or more request *Aura* objects which define its scope of interest (as in the spatial model). These are also created and managed by the local *World* proxy object and are stored in a second spatial octree (see figure 8.5 on page 118).

Fig.8.5: spatial trading based on auras in a single process

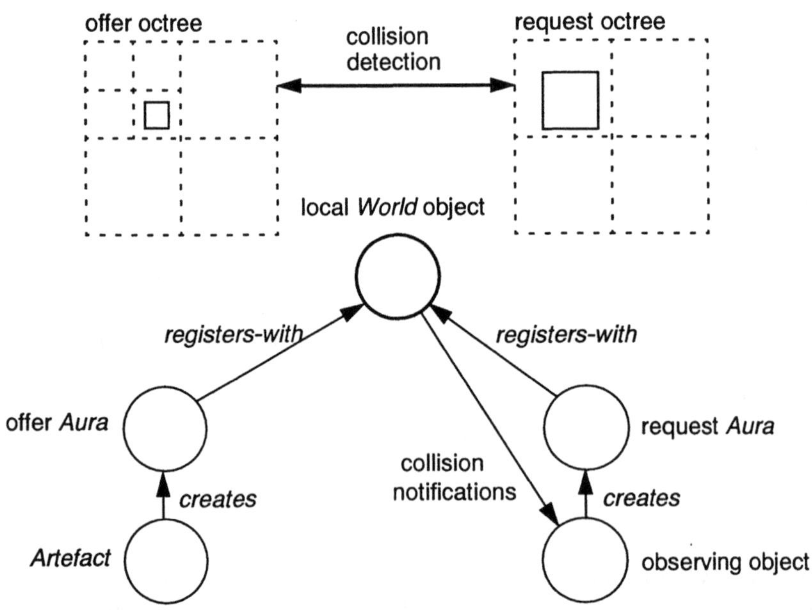

Each *World* object (master or proxy) keeps track of all local *Aura*s for that world in its two octrees and performs incremental collision detection between the offers and the requests. Spatial trading is performed locally in each process and depends on prior replication of remote artefacts (performed by *Group* objects and described in the next section). It is important for the realisation of spatial trading that this collision detection supports incremental collision tests, i.e. identifying only *changes* in aura collision due to possibly small movements and changes of auras. This gives an efficient realisation of two of the distinctive aspects of spatial trading compared to more traditional attribute-based trading (see chapter 6): both offers and requests have persistent representations within the spatial trading service; and offers and requests can be incrementally modified.

The local *World* object uses a callback mechanism to keep each request *Aura*'s owner up to date with all overlapping offer *Auras*. Normally offer *Auras* are dimensioned so as to completely enclose the corresponding artefact: this allows aura collision to take account of the differences of size between artefacts (e.g. the difference in size between an embodiment and a building). Request auras may be any size and they normally represent an upper bound on the extent of the corresponding focus.

To summarise the structure so far (as in figure 8.3 on page 114), a *World* object represents a virtual world and is the initial point of contact on joining a virtual world. This is replicated locally in any process which joins the world. *Artefact* objects represent artefacts within the virtual world such as rooms, parts of a visualisation or the embodiments of users or agents. These are replicated on demand by the *Group* objects associated with worlds and third party objects (the realisation of third party objects is described in the next two chapters). Every *Artefact* has an offer *Aura* which is managed by the local *World* object; this performs spatial trading between offer and request *Auras* to allow objects and applications to discover and observe *Artefacts* within the shared world.

Group

This section describes how *Artefacts* are replicated in processes which have joined a common world. This is handled by *Group* objects. The *mechanism* of replication is described here; details of the *management* of replication are deferred to chapter 10.

As described above, each artefact is represented by a single master object of class *Artefact* which exists in the process in which the artefact was created. This is the object which must be manipulated (and communicated with) to change the appearance of the artefact in the virtual world or to generate medium-specific events. *Group* objects associated with the virtual world (as introduced above) and with the third party objects within it (described in chapter 10) handle the replication of *Artefacts* in other processes which have joined the same world. Each local replica or proxy is an instance of a cut-down artefact class (called *CveArtefactProxy*) which allows querying of attribute values and handling of callbacks but no updating or modification. As with *World* proxies (described above) this proxy acts on behalf of the master *Artefact* object within a particular process. For example, each *Artefact* proxy maintains a local *Aura* so that spatial trading can be performed within each process as described above.

The granularity with which replication is performed is one medium of one *Group* object, i.e. one medium of the artefacts associated with one *Group* object. Replication is performed on demand. This is determined by the *Group* object which monitors local request auras and takes into account the effect of third party objects as described in chapter 10. The overall result is that (at least) all remote artefacts within aura range of a local request aura (subject to the effects of third party objects) will be replicated locally for all media of interest (as specified in the request aura(s)). The next section describes the pattern of interaction and communication which creates and maintains these artefact replicas.

Replication and Communication

Figure 8.6 on page 121 shows the details of the replication process for the *Group* associated with the world as a whole. Recall that *Group* objects are themselves replicated prior to the *Artefact* replication considered here.

Initially, only process 1 exists (a world server, for example). This contains the *World* master object and the associated *Group* master object. Associated with each *Group* are one or more pairs of multicast groups. When a new *Artefact* is created within this process it learns from the *World* object the identity of the *Group* object and the associated multicast groups (1) (one "sender" and one "update" group are shown in the figure; there may be others for other media). The *Artefact* then joins the *sender* group (2a). This marks this *Artefact* as being associated with this *Group* for the purposes of replication (it is used as an implicit membership list). The *Artefact* then multicasts its initial state to the *update* group (2b). If any processes are already replicating this *Group* and medium then this allows them to create a new local proxy for this *Artefact.* Alternatively, suppose that another process subsequently wishes to replicate this *Group*. For example, a user client may enter the world (such as process 2 in the figure). The *Group* proxy in the new process determines from the local request *Auras* that replication is required. It then joins the corresponding *update* multicast group(s) so that it will be able to keep any *Artefact* replicas up to date (3a). It then asks its own master object for the current states of the *Artefacts* in the *Group* (3b). The master object relays this request to the corresponding *sender* multicast group (3c) so that it is received by all of the master *Artefacts* associated with the *Group* (3d) (the master *Artefact* joined this multicast group at (2)). When the master *Artefact(s)* receive this request for state they respond by sending a copy of their current state in that medium directly to the requesting *Group* proxy object (4). The *Group* proxy can then create the local *Artefact* proxy objects (5). Note that these massage sends are all reliable.

Whenever an *Artefact* master object is changed (or emits an event or message) it multicasts this to the *update* group (6a). This will now be received by the *Group* proxy (6b) and used to keep the local *Artefact* proxy up to date (6c).

The figure shows the *Artefact* master object and the *Group* master object in the same process but replication works in the same way when they are in different processes. *Group* master and proxy objects operate in the same way with regard to replication (except that the master object is the only one which sends directly to the *sender* multicast group). Note also that each *Group* may be associated with a number of sender and update multicast groups, each handling one or more media; the number and allocation of multicast groups is determined by the *World* object (on a per-world basis).

Fig.8.6: artefact replication by *Group* objects

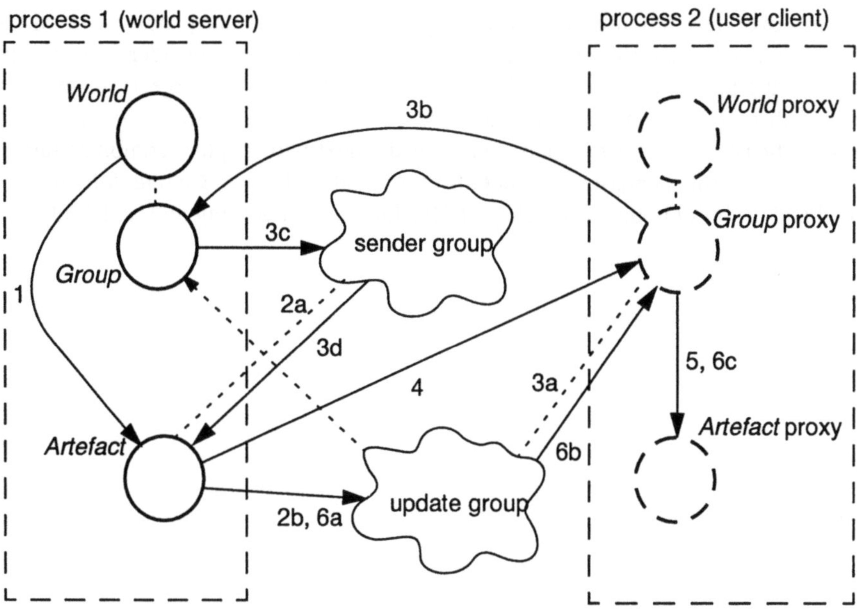

1. An *Artefact* learns from the *World* the identity of the *Group* object and its associated multicast groups.
2. The master *Artefact* joins the sender multicast group and multicasts its initial state to the update group.
3. A *Group* proxy that wishes to replicate this group joins the appropriate update multicast group and passes a request for initial state to the *Group* master which in turn multicasts it to the appropriate sender multicast group so that it is received by each master *Artefact* associated with the *Group*.
4. Those *Artefacts* send their current state directly to the new *Group* proxy.
5. The *Group* proxy then creates initial *Artefact* proxies.
6. When a master *Artefact* is changed it sends update messages to the update multicast group which are received by the *Group* proxy which keeps the corresponding *Artefact* proxy up to date.

Summary

Referring back to figure 8.3 on page 114 the normal operation of MASSIVE-2 in the absence of third party objects should now be apparent. A *World* object represents a virtual world and provides an initial point of access. This may be found using the normal (non-spatial) trading facilities of the node manager(s). A world may be populated by *Artefact* objects which may have representations in one or more media (spatial, text, 2D graphics, 3D graphics and audio). Each *Artefact* has an offer *Aura* which is used in spatial trading. Request *Auras* are created by observing objects such as the

User object which manages interaction with a normal participant. On-demand replication of *Artefacts* is handled by the *Group* objects.

This overview has not considered any aspect of third party objects in MASSIVE-2: this is the subject of the next two chapters. Chapter 9 considers the calculation and application of awareness in the presence of third party objects. Chapter 10 deals with replication management and the coordination of multiple *Group* objects (and multicast groups). As in part I, each of the next two chapters combines further details of the implementation with details of use and evaluation. Chapter 9 focuses on social and usability issues while chapter 10 addresses networking and scalability.

9. Contextualised Awareness

The previous two chapters have presented the third party object extension to the spatial model of interaction and the MASSIVE-2 CVE system which demonstrates it. Chapter 5 in part I described and evaluated computational support for awareness as proposed in the spatial model and as implemented in MASSIVE-1. This chapter extends that analysis to consider contextualised awareness as it arises in the spatial model with third party objects as presented in this thesis. As in chapter 5 the focus of the evaluation in this chapter is on awareness as it impacts the user. The background for this is the same set of sociological motivations which lay behind the original spatial model work, i.e. the emphasis is on open, flexible, malleable and individual control of interaction.

Section 9.1 describes the implementation and capabilities of the awareness negotiation process including third party objects in MASSIVE-2. This demonstrates the feasibility of implementing an awareness system based on the spatial model and third party objects. It also indicates the scope and capabilities of the implementation and may be used to inform other such implementations. Section 9.2 describes the use which has been made of the system and which forms the basis for the reflection and evaluation of section 9.3. Because MASSIVE-2 is somewhat newer than MASSIVE-1 its use has been correspondingly limited. Consequently more of the use and evaluation sections focus on demonstrations and example applications rather than use in trials (as was the case in chapter 5). However the system has already seen significant use on a number of occasions and is being used and developed in a number of recent projects (listed in section 9.2.5).

9.1. Implementation

This section describes MASSIVE-2's implementation of the spatial model of interaction and third party objects, focusing on the calculation of awareness and the realisation of third party effects. Consideration of distribution and replication issues is deferred to chapter 10. Section 9.1.1 begins by describing the realisation of focus and nimbus and the overall framework within which awareness is calculated (c.f. section

5.1's description of MASSIVE-1). Section 9.1.2 then explains how third party objects are implemented and how their effects are integrated into the awareness calculation process. Finally section 9.1.3 describes the additional features and facilities which support the realisation of abstractions (i.e. secondary source effects).

9.1.1. Focus, Nimbus and Awareness

In order to make effective use of multicast communication MASSIVE-2 avoids the connection-oriented peer-to-peer negotiation of awareness used in MASSIVE-1 (see section 5.1). It was found that in practice the prior system made universal use of standard functions for calculating focus and nimbus and for combining them to yield an awareness value. Consequently one party *could* have performed the full awareness calculation provided it knew the parameters which the other was using for the focus and nimbus functions. This is precisely what is done to calculate focus, nimbus and awareness in MASSIVE-2.

A standard function is used when calculating focus and nimbus and multiplication is used to combine these values to yield awareness (as in MASSIVE-1). The focus/nimbus function is extremely close to that in section 5.1.2 (see figure 5.2 on page 59). Each artefact's attributes include the parameters which it wishes to use for calculating focus and nimbus in each medium. These attributes are replicated when the artefact is replicated (for details of the replication process see section 8.4 and chapter 10). These attributes are used to calculate focus, nimbus and awareness locally. The awareness calculation process has (in the absence of third party objects) essentially the same functionality and expressiveness as awareness calculation in MASSIVE-1. It can also be extended to support other functions for focus and nimbus if a mobile code facility is available to allow distribution of alternative *functions* rather than just parameters. This facility could be provided using Java, which has recently been introduced as an embedded language in MASSIVE-2 using the Java 1.1 Native Interface.

Compared to MASSIVE-1 there have been two modifications to the interpretation and calculation of focus and nimbus (which also affect other partial awareness values such as the effects of third party objects, described below). First, MASSIVE-2 allows values of focus and nimbus which are greater than one; this was not permitted in the previous system. So values of focus greater than one can now be used to compensate for low values of nimbus. For example, an observer can "magnify" a whisper up to normal awareness levels. In some situations this behaviour is correct, for example a low nimbus value may simply indicate a non-specific source of information (i.e. not specifically directed at the listener). However in other situations a low nimbus value may indicate a desire for (partial) secrecy. For example, someone may wish an observer to know that they are involved in a side conversation but may not wish the observer to hear what they are actually saying. To support this distinction between degrees of projection on the one hand and security on the other MASSIVE-2 divides nimbus into two components: a nominal or suggested value (as before) and also a maximum value. An observer can increase the nominal value by increasing their focus but they cannot change the maximum value. So the maximum value limits the final awareness which may result and so can be used to enforce restricted awareness even with focus values much greater than one. The nominal and maximum values are calculated using the same standard focus/nimbus function applied to their own further medium-specific sets of parameter values.

The next section describes how third party objects are introduced into this framework. The details of the awareness calculation process are also described after third party objects have been introduced.

9.1.2. Third Party Objects

This section describes the realisation and capabilities of third party objects in MASSIVE-2 in five stages. These are: representation of third party objects; third party activation; direct awareness relationships; secondary sourcing of information (support for abstractions); and combining multiple third party objects. These are dealt with in turn below. Third party objects also play an essential role in the management of artefact replication, however discussion of this is deferred until the next chapter.

Representing Third Party Objects

In the extended spatial model any object can in principle act as a third party with respect to other objects. This is the same in MASSIVE-2: any artefact can act as a third party object. The *Group* class which handles replication (see section 8.4) is also responsible for representing and managing the details of third party objects. The master *Artefact* class includes a method to specify a third party role and effects for the artefact. When this method is invoked the master object creates for itself a new master *Group* object and configures it as specified. This *Group* object's potential role in artefact replication is described in chapter 10. Each replica (proxy) of the *Artefact* also creates a local proxy of the new *Group* object, but this not used in awareness calculations (it is only used for artefact replication). Third party awareness effects are controlled by the *Group* master object. The following sections describe how this *Group* object realises the activation and effects of the third party object.

Third Party Activation

MASSIVE-2 supports a significant and flexible subset of the third party object concept. This subset was described in terms of its presentation to the user in section 8.2. MASSIVE-2 supports two patterns of third party activation: membership and hybrid (defined in section 7.1.2). In each case the system supports membership based on spatial containment. These third party objects are referred to in this thesis as *regional* third party objects or simply *regions*. In the general model, third party membership may be based on any attributes of the objects concerned, giving a more general notion of groups. However within the context of this thesis resources have been concentrated on the more directly "spatial" facilities of the extended model. This is consistent with the emphasis throughout the prototyping work on the potential affordances of 3D Cartesian spaces, which are made tangible and manipulable in a graphical CVE.

Third party objects which are activated based on common membership consider the membership of both objects while third party objects based on hybrid awareness consider the membership of only one of the objects. The region of membership of a third party object in MASSIVE-2 is exactly the same as the artefact's normal spatial extent. The membership region also changes position and size as the artefact moves and changes size. Consequently regions can be both mobile and variable in size. Every artefact which is wholly contained within a region is considered to be a member of that region. The master *Group* object for the region has a request aura which it uses to identify *Artefacts* as they enter and leave its region of membership (refer to section 8.4.2, *Aura*, for a description of offer and request *Auras*). When the *Group* object observes an *Artefact* entering its region of membership it sends a message to the *Artefact* master object. This message specifies the identity of the region and the details of its effects (which are described in the next section). Similarly, when the artefact leaves the region the *Group* object informs it of this.

Fig.9.1: information flow for an Artefact *A* entering a third party object *R*'s region of membership

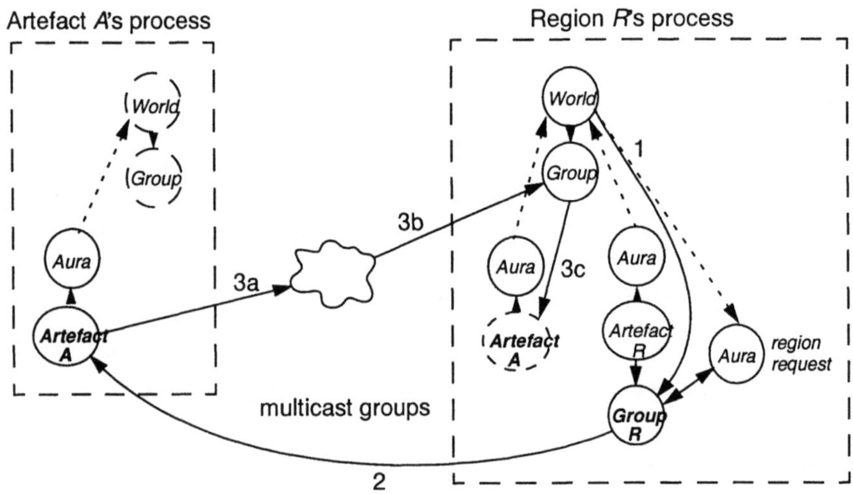

1. The region *R*'s *Group* object discovers through an aura collision that the *Artefact* should be a member of the region.
2. The *Group* object notifies the master *Artefact A* of this; the master *Artefact A* updates its attributes to reflect its membership of this region.
3. The change in the *Artefact*'s region membership attribute is distributed to each proxy via the appropriate update multicast group and the *Group* objects associated with it.

Each *Artefact* uses these messages to maintain a publicly visible attribute which specifies the regions to which it currently belongs and the effects of these regions.

This information is replicated in the spatial medium and so is available to all *Artefact* proxies. This flow of information is illustrated in figure 9.1 on page 126 which shows a remotely mastered *Artefact A* entering a third party object *R's* region of membership.

The next section describes how this information is used when calculating awareness for existing awareness relationships.

Existing Awareness Relationships

Section 9.1.1 introduced the basis of awareness calculation in MASSIVE-2. This section describes the details of this process and how it accommodates the adaptation effects of third party objects.

In MASSIVE-2 a third party object's effect on an existing awareness relationship is specified by:

- a factor by which the nominal awareness is multiplied, giving amplification or attenuation of awareness;
- a maximum awareness value, equivalent to the maximum nimbus value described in section 9.1.1, which imposes a limit on the final level of awareness (for security or privacy); and
- an optional awareness threshold value which is used for group level of detail regions as described below.

Consider the case of a *User* object wishing to know what awareness the user's embodiment, *A*, has of another artefact, *B*. Being an object-oriented system the procedure for finding artefact *A*'s awareness of artefact *B* is to ask it, as follows.

- *A* asks *B* to evaluate its current nimbus on *A*. *B* (or its local proxy) receives this request and feeds *A*'s location and orientation together with its own location, orientation, size and medium-specific nimbus parameters into the standard function to find the value of its nominal and maximum nimbus on *A*. It returns these values to *A*.
- *A* queries *B* for it's current location and size and feeds these, together with its own location, orientation, size and medium-specific focus parameters into the standard function to find its focus on *B*.
- Artefact *A* then asks *B* to calculate the joint effect of third party objects on their awareness and gives *B* the details of *A*'s current region memberships (see detail below). B returns an overall multiplying factor and overall maximum value.
- *A* multiplies the returned nominal nimbus value with its own focus value and the third party factor to give its nominal awareness of *B*. *A* then compares this nominal value to *B*'s specified maximum value and the third party maximum value and returns the smallest value.

For each third party object one of four conditions must hold according to the memberships of *A* and *B*: neither artefact may be a member of the region; artefact *B* alone may be a member of the region; artefact *A* alone may be a member of the region; or both artefacts may be members of the region. The information which *A* and *B* have about their region memberships specifies the effect of each region for each of these joint membership conditions listed above. Each effect has the components listed above and may be medium specific. As an example table 9.1 on page 128 shows the different effects of a crowd third party as a function of the (non-)membership of the observer and the observed. The only significant effect occurs when the observer is not a member but the observed is in which case the crowd suppresses direct awareness (the observer would see only the crowd abstraction). In this case (with *B* a member but *A* not) the maximum awareness returned to *A* after *B*'s calculation of third party effects would be zero and so *A* would return an overall awareness of *B* which was zero.

Table 9.1: effects of a crowd region according to membership

	Observed member	**Observed non-member**
Observer member	no effect, normal awareness	no effect, normal awareness
Observer non-member	maximum = 0, awareness suppressed	not considered, no effect

If a threshold awareness value is specified by a third party object then the observing artefact's (*A*'s) awareness of the third party artefact itself is calculated (using the full awareness calculation process). If this awareness value exceeds the specified threshold value then calculation of *A*'s awareness of *B* proceeds. However if *A*'s awareness of the third party artefact is lower than the threshold value then *A*'s awareness of *B* is suppressed (i.e. set to zero) and no further calculation is required. This threshold test is used for group level of detail management and ensures that an artefact which has insufficient awareness of a level of detail region is not aware of the region's members.

This process is repeated for each region in turn in a consistent but otherwise arbitrary order. Once all of the regions have been considered the final nominal and maximum nimbus/awareness values are returned to *A* which completes the calculation as before.

Secondary Sources

The previous section has described how adaptation of existing awareness relationships is performed in MASSIVE-2. The other key effect of third party objects is to provide indirect information about other artefacts, i.e. secondary sourcing. This is dealt with in this section.

In one sense the implementation of secondary sourcing is entirely trivial: one artefact redistributes some or all of the information which it receives from another set of artefacts. This could be realised in any system without awareness or third party objects. However MASSIVE-2 and third party objects provide additional support for secondary sources in a number of areas listed below.

- Secondary sources are linked to region third party objects, thereby integrating regional adaptation as discussed below.
- In MASSIVE-2 artefacts only become members of a region when they are fully contained by the region. Consequently the spatial extent of the region also bounds the scope of the secondary source in a consistent and intuitive manner.
- The adaptation effects of a region can provide awareness-driven selection between a secondary source and the individual artefacts which are its members (the threshold adaptation effect described in the previous section suppresses direct awareness relationships when awareness of the region falls below some critical value).
- There is also an optional threshold component in nimbus (which is not used in normal awareness calculation) which is used to indicate to observers that the third party object should not be rendered when direct awareness of it exceeds the specified value. This is the complement of the direct awareness threshold. So at lower awareness values only the secondary source will be seen. At higher awareness values the region members will be seen directly while presentation of the secondary source will be suppressed.
- A region can indicate that its members should not render it. This can be important to avoid feedback or visual interference to its members from the secondary sourced information.
- A standard class is provided which implements many of the basic elements of secondary sourcing. Application programmers can use the simple abstractions already provided, or can override the abstraction and presentation methods to create new types of secondary sources.

In a general system, when providing a framework for creating secondary sources and abstractions, the main challenge is managing the various flows of direct and indirect information. In particular, care must be taken (and facilities must be provided) to avoid feedback loops and to resolve potential conflicts between contradictory or delayed versions of the same underlying information. The awareness based framework of the spatial model with the inclusion of third-party adaptation is an effective way of doing this.

Multiple Third Party Objects

In general there may be any number of co-existing regions and secondary sources. This is handled naturally and automatically by MASSIVE-2 because of the two factors listed below.

- Every region is also a normal artefact and as such can be a member of another region or regions. Consequently hierarchies of regions will form automatically, based on the spatial containment rule used to establish region membership. As regions move and change size these hierarchies will adjust and reform accordingly.
- Regional adaptation effects combine unambiguously when artefacts are members of multiple regions because the operations used (multiplication and minimum) are separately commutative and associative.

The main area which requires additional support when multiple regions are involved is that of secondary sourcing and of abstraction in particular. Specifically the possibility of nested abstractions requires that an outer secondary source abstract over nested abstractions. This may require additional meta-information relating to the abstraction process. For example, an abstract view of a crowd region may be relatively ambiguous. However, creating an abstract crowd-of-crowds would require explicit information about the number of people in each of the sub-crowds. This type of information is not normally available in artefacts and requires additional explicit support. A possible area of future work would be to establish a core set information which would facilitate effective nesting of a broad range of abstractions.

This is also a medium-specific issue as demonstrated by audio abstractions. MASSIVE-2 provides simple audio abstractions based on mixing and optional audio processing (see section 9.2 for examples). Because of the nature of the audio medium this requires no additional information, and arbitrary hierarchies of audio abstracting regions will work correctly in the current system given sufficient CPU power.

The current system supports flexible spatially organised hierarchies of regions and abstractions. The next section describes a number of applications and demonstrations of third party objects which make use of multiple and even nested third parties.

9.2. Use

The previous section described how the awareness effects of third party objects are implemented in MASSIVE-2. Before the reflections and evaluation of section 9.3 this section describes a number of existing applications and demonstrations of MASSIVE-2 (including third party objects). These are: the "MASSIVE" mixed reality poetry performance; the Arena; the Panopticon Plaza; and the WWW-3D collaborative World Wide Web browser. A final section lists other use of the system.

9.2.1. Poetry

The most demanding use of MASSIVE-2 to date was the staging of a mixed reality Hip-Hop poetry performance called "MASSIVE" as part of the NOWninety6 arts festival in Nottingham (Sunday 10th November 1996) (see [Benford et al., 1997] for further description of this event). Figure 9.2 on page 131 shows the overall organisation of the venue. Four poets took turns to perform at the front of the auditorium (a cinema) to a live audience. The poets wore magnetic trackers on their hands and head which drove a representation within the virtual world. A view of the poet in the virtual world was projected onto the cinema screen. Ten members of the public at a time used the workstations in the bar area to share the virtual world and to view the performance. These audience members also had microphone headsets which allowed them to hear the poet, to talk to one another and to be heard by the live audience (at the manager's discretion). Two non-graphical workstations (the "manager" and the "trouble shooter") were used to coordinate and monitor the event. All of the machines were connected using a dedicated 10 Mbit Ethernet. A total of over 60 members of the public participated as virtual audience members during six organised sessions running over three hours of performance and free interaction.

Fig.9.2: layout of the "MASSIVE" event of NOWninety6

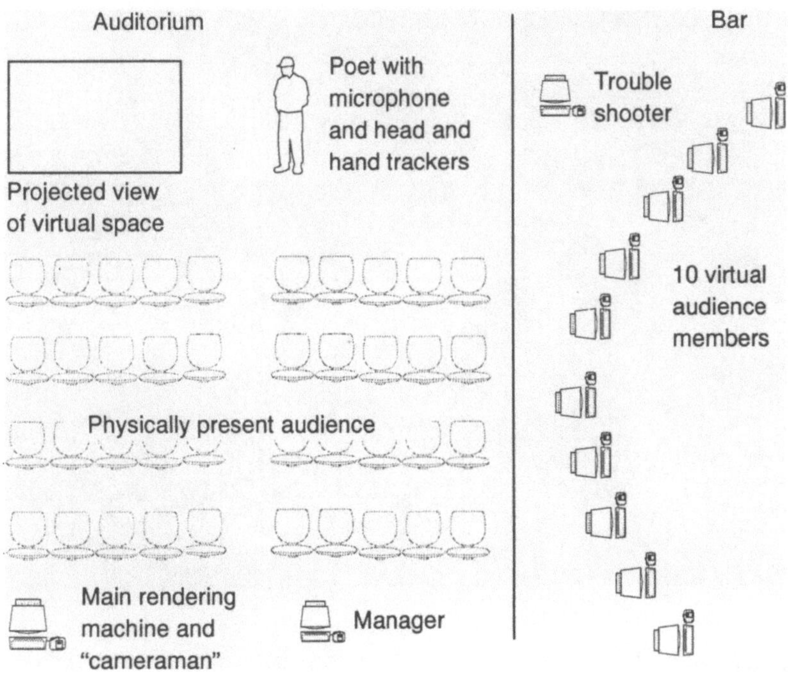

The virtual performance space is shown in overview in figure 9.3 on page 132. The poets performed in the central stage area. Around this central performance space were four outlying regions, each dedicated to a different poet and including lines from their work. Each of these regions was realised as a level of detail region: when viewed with low awareness they appeared as featureless coloured cones (as in the figure). However when viewed with high awareness (i.e. from nearby or within) the internal structure and any occupants became visible, audible, etc. Participants inside each region were also able to hear and see out of the region (e.g. to see when the performance began). This spatial structuring was designed to encourage exploration and permit the formation of local conversational groups within individual regions (no secondary sourcing is present). The poetry world and performer interface were created by Dr. Dave Snowdon of the University of Nottingham using geometries by Mr. Sean Varney, a virtual reality artist, and using the standard world and region specification facilities of MASSIVE-2.

Fig.9.3: overview of the "MASSIVE" world with outlying regions

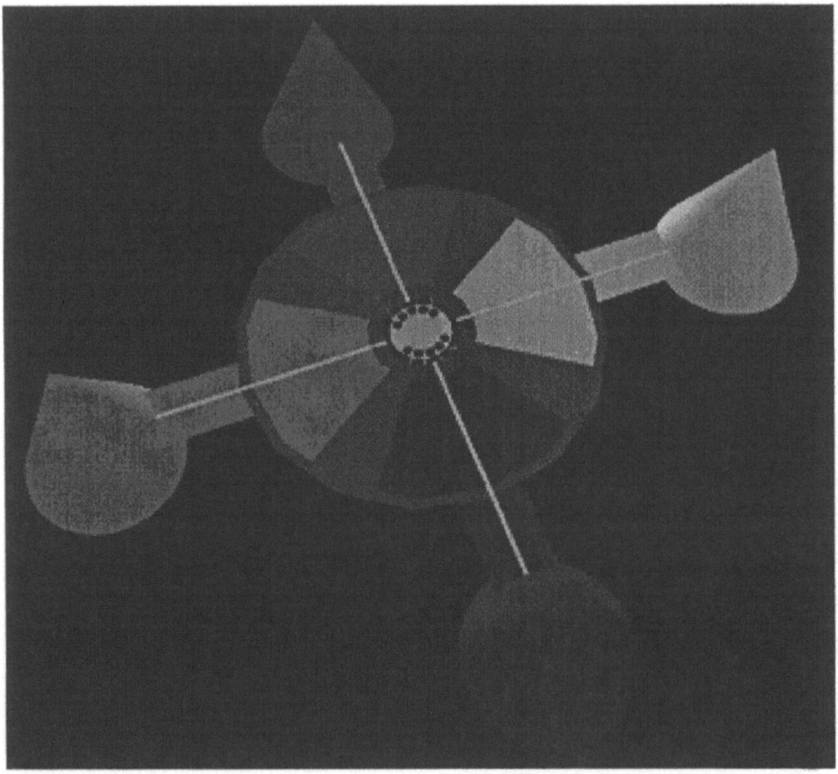

9.2.2. The Arena

The original demonstration application for MASSIVE-2 is another virtual performance space, but more structured than the poetry world. The structure of the Arena is shown in figure 9.4 (a) on page 133. The Arena itself is a virtual building for hosting performances and events and is realised as a bidirectionally closed region. This prevents those within the Arena being distracted or overheard by those outside. Within the Arena is a central performance space with two fixed crowd regions on either side. Each crowd region allows its members to observe both the performance space and other members of the same crowd with normal awareness. However it presents external observers (i.e. the performers and the members of the other crowd) with only an abstracted and secondary sourced view of its members. This abstraction comprises a simple graphical embodiment which changes size to indicate the current number of members and a combined audio signal which is mixed-down from its members' individual audio streams and low-pass filtered (to emphasise its indirect nature). Figure 9.4 (b) on page 133 shows a view from within one crowd, looking over the performance space to the opposing crowd abstraction (the "performers" shown in the figure are embodiments from the poetry event).

Fig.9.4: the Arena

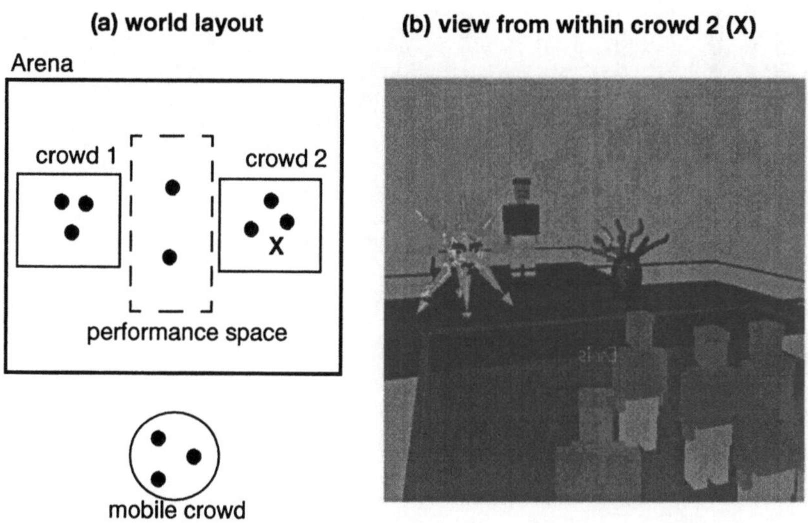

The Arena also includes a mobile crowd region at the exit which follows its members as they move about. In principle, crowds or crowd generators could be placed in a virtual world where crowds are expected to form (for example after a performance). These could reduce the system requirements related to the high density of participants by introducing secondary sourced abstractions. The crowd might then be garbage collected when it was no longer useful (e.g. when it had too few members). The Arena application is populated by simple simulated users which move through the world and emit audio samples.

The Arena demonstrates nested regions (the crowds within the Arena), secondary sourcing and mobile regions.

9.2.3. The Panopticon Plaza

The Panopticon Plaza is inspired by Jeremy Bentham's ideas for a panopticon prison as described by Foucault [1977] which proposes a cunning arrangement of walls and openings to make the inmates visible to the guards but without being able to see them in return. The centre-piece of the Panopticon Plaza, which is shown in figure 9.5 on page 134, is a panopticon cell. This is a region third party which suppresses any awareness which its members might have of external participants and objects. It has no effect in the other direction, allowing external observers full visibility, audibility, etc. of the cell's occupants.

Fig.9.5: the Panopticon Plaza

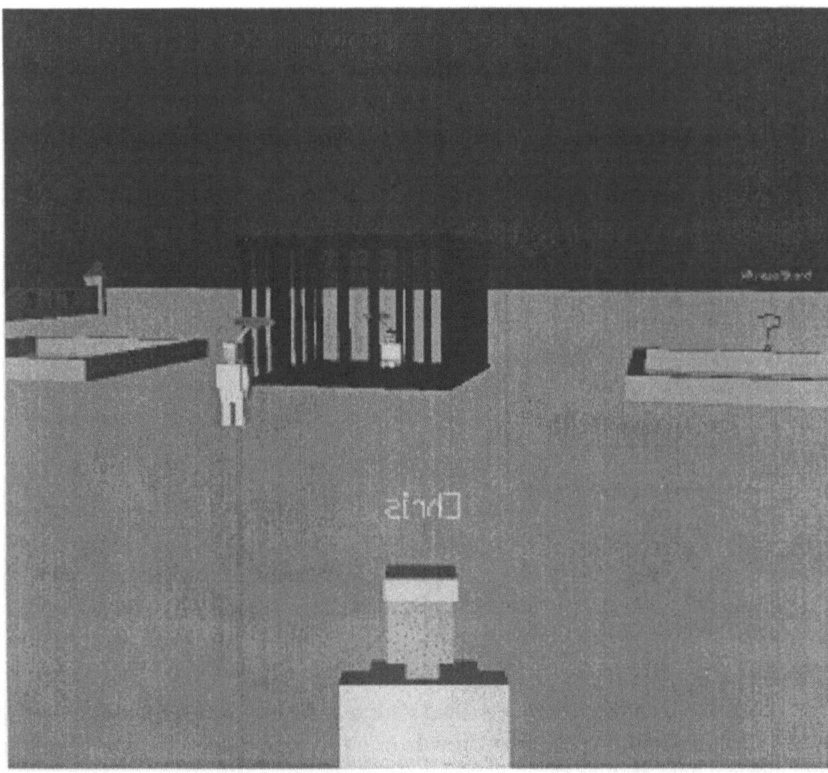

On either side of the panopticon cell are two other regions which extend the legal theme: a witness stand and a jury box. Each of these is also a unique third party region. The witness stand allows the witness to see and hear what is happening outside, but prevents those outside from seeing or hearing the witness directly. It does however provide a secondary source for the witness's audio, but this is subject to signal processing before re-transmission to disguise the witness's voice and so preserve their anonymity. The jury box also preserves the anonymity of its members while allowing them to see and hear what is happening outside. The jury box provides no secondary sourced audio but it does provide a graphical abstraction; this shows the current state of the jury's voting as totals for "yes", "no" and "undecided". Normal users can also join the jury by entering the jury box. A specialised window in the standard user interface allows them to register a vote. The jury box preserves both the anonymity of the jury members and the secrecy of the ballot - only the jury box itself "sees" what each individual is voting.

The Panopticon Plaza shows that third party objects can be used to realise useful effects which would be difficult or impossible in the everyday world.

9.2.4. WWW-3D

The final example application is "WWW-3D" which has been created using MASSIVE-2 by Marten Stenius from the Swedish Institute of Computer Science and Dr. David Snowdon from Nottingham University (see [Snowdon et al., 1997] for details). WWW-3D is a novel Web browser which displays both the linked structure of collections of pages and the actual content of those pages within a combined 3D graphical view.

Fig.9.6: the WWW-3D graphical Web browser

(a) view from within a page (DIVE) **(b) view from a server region (MASSIVE-2)**

Figure 9.6 (a) on page 135 shows the view from within one WWW page: the page's contents are spread around the surface of a sphere in the foreground while links stretch off into the distance to other pages (which are shown from the outside as opaque spheres). This application was originally implemented using the DIVE system [Hagsand, 1996] but has been ported to MASSIVE-2 and extended to take advantage of the provision of third party objects. Specifically, the MASSIVE-2 version now groups WWW pages according to the server which hosts them and all of the pages on a single server are enclosed by a level of detail region. From outside of a server region it appears as a single large wire-frame sphere. The colour of the sphere

gives a clue as to the amount of activity occurring around those pages. On entering a server region all of its pages (and the other participants who are using them) become directly visible while other server regions are still visible in the distance. Figure 9.6 (b) on page 135 shows the view from within one server region looking past the local pages to two other servers.

WWW-3D shows that regions can be used to create "data districts" in large-scale data visualisations. The activity indication for each server region also demonstrates another application of secondary sourcing and abstraction.

9.2.5. Other Use

MASSIVE-2 has also been used in a number of other situations listed below.

- Like MASSIVE-1 it has been used on about 5 occasions to hold the weekly laboratory meetings of the Communications Research Group (with six to eight participants).
- MASSIVE-2 has been used in wide-area trials over the UK's SuperJANET educational network between Nottingham and Reading Universities within the HIVE (large sale real time multi-user virtual reality) project funded by the UK's Engineering and Physical Sciences Research Council.
- It is currently being used and extended by three PhD students in the group: Mike Fraser is exploring object-centred interaction (such as common-focus third party objects); Gail Reynard is using it as a test-bed for awareness-driven video streams in CVEs; Ian Taylor is looking at the visualisation of relational databases.
- MASSIVE-2 has been used for numerous demonstrations in the laboratory.
- It is being used at British Telecommunications Laboratories as an experimental application for generating dynamic network resource reservations (using RSVP [Braden et al., 1996]).
- MASSIVE-2 has recently been ported to the Windows NT operating system in anticipation of larger-scale usage, for example in the JTAP funded "Virtual Campus" project which aims to make CVE technologies available to the UK postgraduate researcher community.

9.3. Evaluation

This section presents an evaluation of MASSIVE-2 and the third party object concept which it demonstrates. It has not been possible to perform formal usability evaluations within the scope of this work and so these results are necessarily preliminary. The system has already had significant use in its own right (see section 9.2) and this section also builds on experience gained with MASSIVE-1, as presented in section 5.3. This section is divided into three main parts: success; issues of use and visibility; and issues of implementation and capabilities.

9.3.1. Success

MASSIVE-2 has been successful in a number of respects and this reflects well on the proposed extensions to the spatial model of interaction which were presented in chapter 7. All of the objectives identified for part II of this thesis (in section 6.4.3) have been met: all interaction is controlled by the spatial model of interaction and third party objects; awareness negotiation includes contextual factors - through the activities of third party objects; and network multicasting has been employed for the core distribution services.

The system also *works*, and so demonstrates that at least a subset of the model can be implemented with acceptable overheads. This subset includes both adaptation and abstraction with activation based on membership and hybrid awareness. This has been implemented in the context of multicast based communication and in addition to the basic spatial model facilities of aura, focus, nimbus and awareness. The implementation style is very different to that used in MASSIVE-1 and demonstrates that the components of the spatial model can be realised in a range of system architectures.

As described in section 9.2, MASSIVE-2 also shows that third party objects can be used to create a range of useful contextual effects. These include closed rooms, open zones or cells, level-of-detail groups and crowds. Furthermore, these can be combined into spatially defined hierarchies such as crowds within a building within a zone, etc. Any region can be independent and mobile and take its effects with it. Also, although not demonstrated in the example applications, third party objects can be introduced into a running system and will fit themselves into the region hierarchy. In addition, a number of more unusual effects and affordances for interaction have been demonstrated including the panopticon cell, the witness stand and the jury box of the Panopticon Plaza demonstration (section 9.2.3).

Secondary sourcing and abstraction have been demonstrated in a number of contexts in the 2D and 3D graphical and the audio media. The crowd demonstrates abstraction in all of these media. WWW-3D also demonstrates less direct abstraction in the context of information visualisation. Both of these examples reduce the amount of visual clutter, graphical and audio rendering and potentially reduce network and computation requirements (in conjunction with management of replication, discussed in the next chapter). At the same time, secondary sourcing provides a degree of awareness of the activity and occupancy of a region. This is one of the most interesting and promising uses of third party objects. In particular, the use of abstraction and secondary sourcing for the audio medium appears to be particularly effective, allowing multiple audio streams to be merged and managed in a contextually sensitive way.

Finally, MASSIVE-2 also demonstrates the potential of secondary sourcing to provide other effects including cross-medium adaptation (the jury box voting abstraction) and support for anonymity (the witness stand's audio processing). The same facilities might also be used to manage interaction in a more formal way such as floor control.

9.3.2. Visibility and Usability

From chapter 5 one of the key observations of the original spatial model in MASSIVE-1 was that awareness, focus and nimbus were not directly visible to the user. Consequently, users found it difficult to control their interaction appropriately or to understand the reasons behind some presented effects. MASSIVE-2 has not addressed this problem (and was not intended to). Furthermore, third party objects introduce new effects on awareness which are also not directly visible in MASSIVE-2.

Some types of third party object are relatively intuitive. For example, a closed room in which the third party region is aligned with the walls of the room has a natural representation which correlates directly with its effects and makes it easy to understand and work with. Level-of-detail groups and abstractions tend to have distinctive external representations so that it is clear to users whether they are viewing the abstraction (secondary sourced information) or the region's members.

However asymmetric boundaries and regions which have different effects in different media can be unexpected and confusing. For example, when standing in the jury box a user can see and hear everything that is happening outside and it is easy to forget that those outside can neither see nor hear them. There is also no direct feedback that they are not being seen or heard. Similarly, the nature of the panopticon cell is not immediately apparent. It is only by exploring the inside and outside of the region that its effects are revealed. Consequently, either additional cues are needed to indicate the presence and nature of region boundaries (e.g. virtual smears on the "glass") or direct feedback to the user is required concerning ones own visibility and audibility to other participants.

It has also become apparent that it is difficult to create suitable and effective abstracted representations and secondary sourcing effects. The audio medium has been the most successful because simple mixing and filtering suffice in many situations. However the graphical medium has proved particularly difficult given the constraint of actually *abstracting*, i.e. showing less information. For example, how can the *impression* of a crowd of 20 people be created without using a composite geometry made up of 20 embodiments? The current default representation of a crowd grows to indicate the number of members but this can be confusing to novice users and can create problems with visually estimating the sizes and distances of crowds compared to single users in normal environments (see [Benford, Greenhalgh and LLoyd, 1997] for an extended consideration of crowds and medium-specific abstraction in CVEs). Significant further work is needed to explore and identify appropriate representations and forms for abstractions in different media.

9.3.3. Implementation and Capabilities

The previous section considered some areas in which the visibility of third party objects and the spatial model could be improved to increase its usability. There are also a number of areas in which MASSIVE-2 either falls short of the potential capabilities of third party objects or in which further refinements are needed in the detailed realisation of the spatial model within the current framework. These issues are considered in this section.

The main shortcoming of MASSIVE-2 as compared to the third party object model (see chapter 7) is in the area of third party object activation. MASSIVE-2 currently supports activation based on membership of both artefacts or of one only (hybrid awareness). Activation based on common focus is not supported at present. This area is being addressed by Mike Fraser as part of his program of research for a PhD in the area of "object centred interaction"; this combines the common focus aspects of third party objects with observation and analysis of real-world interaction around physical objects. Consequently it is not yet possible to comment on the potential utility and implementability of third party objects activated by common focus.

Activation due to membership in the third party object *model* is defined in terms of the third party object's awareness of another object. In the MASSIVE-2 system this has been restricted to spatial containment within an axis-aligned cuboid bounding region, i.e. effectively, awareness in the spatial medium only and with undifferentiated focus and nimbus. This has sufficed to create simple rooms, buildings, crowds and cells which are (or closely approximate) this form. However it is not sufficient for dealing with more complex patterns of architecture. It also cannot support membership effects with non-spatial components, such as shared interests or security considerations. In theory membership could also depend on logical as well as spatial considerations, e.g. based on an artefact hierarchy. MASSIVE-2 currently lacks any explicit multi-level scene hierarchy and this makes some effects very difficult to achieve, e.g. a number of participants on a moving vehicle or a number of articulated limbs forming a single top-level artefact such as an articulated body. Support for artefact hierarchies would be a particularly useful addition to the system.

Abstraction and secondary sourcing are supported and can be used in MASSIVE-2 as it is at present. However additional support would help in two respects. First, as noted in the previous section, it is difficult to create simple, informative and legible abstractions. This might be addressed to some extent by system extensions. For example, the Projection Aggregation Entity Summary Protocol of [Singhal, 1996] represents the positions of a group or cluster of related entities by a count, average position and spread. The observing machines use this statistical information to synthesise an appropriate and representative scattering of objects without ever communicating all of the constituent positions between the processes concerned. In MASSIVE-2 this would imply the use of another specialised medium or perhaps some form of support for distributed behaviour in order to convey and use specialised abstract representations of this kind.

Abstraction also requires additional support in order to create meaningful hierarchies of abstractions such as crowds within crowds. At present, the top level crowd would see only the sub-crowds and so would give an incorrect abstract representation reflecting the number of *crowds* rather than the total number of crowd *members*. Additional meta-information would be required to support this kind of abstraction of abstractions. For example, each crowd would need to convey to its superior the number of members in a form which the superior crowd could access and use. This meta-information would ideally be of a standard and extensible form.

Finally, further work is still needed to define and refine the choice and interpretation of focus, nimbus, awareness of other effects. It was noted in chapter 5, regarding MASSIVE-1, that natural spatialised interaction does not "just happen" as soon as the spatial model is introduced. It is apparent that, at the very least, careful consideration must be given to the choice of functions for focus and nimbus and to the meanings and effects of different levels of awareness. It must be remembered that the goal is to support natural, flexible and subtle forms of interaction. When is someone *interrupting*? How does their nimbus identify this? How is it reflected in awareness? How is this communicated to the listener? Does it interfere with their other interactions any more than might be expected in the physical world? This remains an area for future work. MASSIVE-1 and MASSIVE-2 demonstrate that a number of facilities and opportunities are achievable, but the full potential and ultimate limitations are not yet apparent.

9.4. Summary

This chapter has considered the implementation, effects and affordances of third party objects in managing and controlling interaction and awareness in CVEs. Section 9.1 described how awareness negotiation and the effects of third party objects are realised in MASSIVE-2. These concrete implementation details show how the extended spatial model might be implementation and illustrate the kind of decisions and additional components which are required to make use of the model's concepts. Section 9.2 then described how the current system has been used. In particular it described a number of applications and demonstrations which are existence proofs of a number of aspects of the model and its implementation. Finally, section 9.3 reflected on these experiences with MASSIVE-2 and provided a preliminary evaluation of the system and the third party object extension to the spatial model.

As in chapter 5, the focus of this chapter has been on the user: how ideal awareness is determined, how information is filtered for presentation, what the system is able to express, what this affords in terms of patterns of interaction. The next chapter provides the more computation and communication oriented complement to this, and might be said to "complete the awareness loop" (of figure 1.3 on page 12) by considering how information exchange is managed and realised in MASSIVE-2 and related systems.

10. Awareness Driven Communication

Chapter 9 has considered how awareness and third party objects are implemented and used in relation to a user of MASSIVE-2. This chapter provides the complementary view of third party objects in relation to network communication. Chapter 6 considered communication and scalability issues in the context of MASSIVE-1 and unicast-based peer-to-peer CVEs. This chapter extends that analysis to encompass a range of systems and approaches and to consider a number of CVE design issues including replication management and the appropriate use of multicast-based networking. This chapter is a key component of this thesis because it demonstrates how third party objects and an explicit computational model of awareness can provide more efficient and appropriate use of multicast communication and network resources that existing ad hoc approaches. The network requirement analysis itself is also a significant component of this thesis.

Section 10.1 describes how third party objects and effects are exploited in MASSIVE-2 to manage multicast network communication (the "awareness driven communication" of the title). This builds on the system overview of section 8.4. Section 10.2 gives some examples of the use and potential benefits of this approach which makes use of multicast group management and secondary sourcing. Section 10.3, which parallels and extends section 6.3, presents an analysis of network communication requirements for MASSIVE-2 and related systems. A model of communication in CVEs is developed and used to compare and relate a range of potential approaches to communication management in CVEs and similar systems. Finally, section 10.4 presents a brief summary and conclusions for this analysis.

10.1. Implementation

Section 8.4 described the overall network software architecture of MASSIVE-2 and the basic replication mechanisms as applied to a single undifferentiated (i.e. fully replicated) world. The previous chapter described how the spatial model of awareness with third party objects was realised within this framework. This section develops these previous descriptions to show how MASSIVE-2 uses third party objects to inform and control partial replication of world content based on awareness, i.e. this section describes the approach adopted to scoping interaction in MASSIVE-2 (see section 2.2.1). This approach is distinguished from contemporary alternatives by its consideration of social factors in interaction (based on the spatial model of interaction and third party objects) and in its expressiveness and flexibility.

Total replication of world state (and the consequent requirement to track all updates) is undesirable since it limits system scalability and potential variation in end-user hardware capabilities because of the inflexible requirements placed on end-user machines (see section 1.2). Total replication is also *unnecessary* with regard to the application's semantics if interaction is based on the spatial model and third party objects. To recap from section 7.3, in the spatial model with third party objects one artefact cannot possibly be aware of another artefact when that artefact is:

- outside of aura range;
- on the other "side" (inside or outside) of an opaque region boundary such as a closed room; or
- inside a group level of detail region (i.e. spatially-defined third party object, as used in MASSIVE-2) when awareness of the region is below the region's threshold value so that only the abstracted view is visible.

In each of these cases replication is unnecessary. MASSIVE-2 avoids artefact replication and corresponding communication in each of these cases, though with certain additional restrictions in the first two cases.

This section describes the implementation of partial replication and multicast group management in MASSIVE-2 in a number of stages. It considers in turn: the units of replication; propagating replica information; obtaining replicas; replication based on aura; replication for opaque regions; replicating for level of detail regions; and replication with region hierarchies.

Units of Replication

The unit or granularity of replication is the smallest unit of information for which replication can be individually controlled. In MASSIVE-1 the unit of replication was a single medium of a single artefact: each process exchanged information on an object by object and medium by medium basis. In MASSIVE-2 a potentially much larger unit of replication is used, namely one medium of one region (and the artefacts which are members of that region). This larger unit of replication is chosen to reduce the number of multicast groups which are required by the system on the basis that they are a potentially limited resource. Multicast group use is analysed in section 10.3.13 where this idea is considered in more detail.

Because multicast communication is being employed it is generally undesirable to send the same update messages to multiple multicast groups. So artefacts which are members of more than one region are considered, for the purposes of replication and communication, to "belong" only to the *smallest* region of which they are a member. If regions are nested within other regions then this is naturally the most deeply nested region. Artefacts which do not belong to any specific region are considered to belong to a "whole world" region, which has no spatial limit. Figure 10.1 (a) on page 143 shows an example world containing artefacts and a number of nested regions. Figure 10.1 (b) on page 143 shows how the same artefacts and regions are organised for replication and communication. For example, a process which chooses to locally replicate the closed room region, *R*, will get local replicas for artefacts *d, e* and *f*. Correspondingly, artefacts *d, e* and *f* all send their update messages, audio packets, etc. to the medium-specific update multicast groups which are associated with region *R*.

Fig.10.1: example virtual world with nested regions

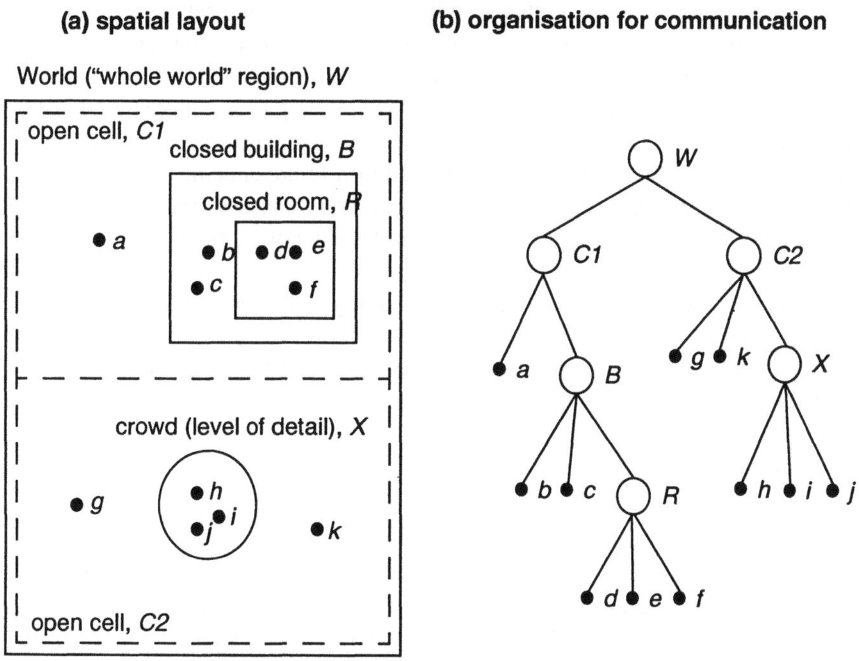

This allocation of artefacts to regions and the propagation of the necessary information is handled by *Group* objects in the same way as regional adaptation (which was described in section 9.1.2 and figure 9.1 on page 126). It is the master *Group* objects which are responsible for managing region membership. When an *Artefact* becomes a member of a region it is informed of the communication channels which are used by the region, i.e. it receives a set of medium-specific multicast addresses. It is also informed of the size of the region, which is updated if the region changes size. This information is in addition to the awareness effects of the region, considered in the previous chapter. Conversely, when an *Artefact* leaves a region it is informed of this in the same way. Each *Artefact* continually assesses its choice of replication region based on its current membership and the sizes of the regions. If an artefact's choice of replication region changes then the artefact will move to the new region.

Propagating Replica Information

As described in section 8.4 when an artefact first joins a world it learns about the *Group* artefact associated with the world as a whole (i.e. the "whole world" region). At the same time it learns the identities of the communication channels which are associated with this *Group* and which should be used when the artefact does not belong to any more specific region.

As an *Artefact* moves about within a world and as regions move and change size the *Artefact* keeps itself a member of the smallest enclosing region and sends any new information (movement, updates, audio, etc.) to the multicast groups associated with that region. It is essential for exploiting the spatial model that group membership for replication corresponds to third party membership for controlling awareness.

To move from one region to another the master *Artefact* announces its departure to the multicast groups associated with the old region and announces its arrival to the groups associated with the new region. In addition to announcing its arrival it also sends a snap-shot of its current state. This is necessary because some processes may be replicating and monitoring the new region but not the old region and would not yet know about the arriving artefact.

This is the one half of the process of managing replication in MASSIVE-2 and might be described as "how artefacts make information and updates available on the network". The second half deals with how processes obtain and make use of this information. This is described in the next section.

Obtaining Replicas

The previous section has described how each artefact monitors its own region membership and sends update messages to the multicast groups associated with its smallest enclosing region. It also notifies each multicast group as it joins and leaves. These messages can then be used to create and maintain artefact replicas.

Section 8.4 has already described how *Artefact* replicas (proxies) are created and maintained in the case of the "whole world" *Group* (see figure 8.6 on page 121). The same procedure is applied to each region-related *Group*, using its own unique set of multicast groups. It was noted in section 8.4 that replication of a medium is only performed as required and that this is inferred from the presence and type of request auras in a given process. The most important part of MASSIVE-2's handling of multicast groups and replication is the development and implementation of this replication policy based not only on aura but also on region membership and awareness.

The description in section 8.4.2, *Groups*, showed how a process in a distributed CVE session could "page in" an individual replication group (i.e. a region), obtaining initial artefact state via a direct request, as well as joining the relevant multicast groups so that these replicas could be kept up to date. The same process is used for regions and is handled by the *Group* objects associated with the region. The last three sub-sections, below, consider the critical management issue of *when* regions should be paged in. There are three forms of replication management implemented in MASSIVE-2 and they related directly to the three points of exploitation of the spatial model extended with third party objects noted in section 7.3 (and repeated at the start of this chapter). These three forms of replication management are: aura based paging; membership paging; and awareness driven paging. They are described in turn.

Aura Based Paging

In the spatial model (both the original form and with third party objects) there can be no direct awareness between two artefacts when there is no aura collision. In MASSIVE-2 an artefact is only invited to become a member of a region when it is fully contained by the region. Therefore the offer aura of any artefact which is a member of a region will always be wholly contained within the offer aura of the region itself. Consequently, if a request aura does not intersect with a region's aura then it cannot intersect with the auras of the region's members. This is actually the weakest of the three limitations of awareness and applies to boundaries which never suppress awareness of their members, i.e. for regions which are open or weakly attenuating (in whatever medium).

In MASSIVE-2 the *Group* object (master or proxy) which is *locally* responsible for an open region monitors request *Auras* in the local process using spatial trading (see section 8.4.2, *Aura*). When there are no overlapping request *Auras* the region is paged out (recall that auras, like paging, are medium specific). When a local request *Aura* collides with the region then the group object pages the region in (obtaining initial state information and joining the update multicast groups). This may occur when an *Aura* is first created or when it is moved or resized or when the region is created, moved or resized. Another *Aura* requesting a different medium would cause that other medium to be paged in independently. Note that every *Aura* implicitly includes a requirement for the spatial medium, since without this the other information (such as graphics and audio) cannot be situated in the space.

This is the first and simplest form of region paging and has no dependence on third party objects. This form of paging is equivalent to the Area of Interest group management proposed for NPSNET-IV [Macedonia et al., 1995]: an *Aura* represents the area of interest and each region corresponds to a single multicast cell. However, unlike that proposal, each region is explicitly represented within the space. This implies additional communication requirements but allows arbitrary sized and positioned regions as well as mobile and dynamically introduced regions.

Because regions have been chosen as the unit of replication this partial replication is still more conservative than is strictly required by the spatial model. Specifically, all members of a region will be paged in if one member might be within aura range. This may be contrasted with MASSIVE-1 in which spatial trading was performed as a central service (rather than locally on each process) and in which replication of artefacts was handled individually using unicast communication. This replication of artefacts over and above the minimum required is an inevitable result of choosing a larger unit of replication (which is motivated by the desire to reduce the number of multicast groups and corresponding management overheads). The overheads of this and other approaches are compared more formally in section 10.3.

Membership Paging

The second form of region paging deals with completely opaque regions as viewed from the outside. Examples include closed buildings, rooms and closed crowds (but not level of detail crowds). A region of this kind uses adaptation to force all non-member's awareness of the region's members to zero. The region may optionally act as a secondary source of information about the region contents but this does not affect paging which is being discussed here. Note that there is no requirement that either members lose or retain awareness of non-members, i.e. members may or may not be able to "see out". For this type of region aura overlap alone (as above) is *necessary* but not *sufficient* to cause awareness between artefacts. As well as aura overlap the local artefact must also become a *member* of the region before it need be paged in.

The local master *Artefacts* cooperate with the region's local *Group* proxy object to achieve this effect. The *Group* object does not perform aura paging as a result of aura overlap if a medium is isolated in this way (note that the same region may have different media paged in different ways because of different adaptation effects in different media). The individual master *Artefacts* notify the local *Group* object when they become members of the corresponding region. Only at this point does the *Group* object page in the appropriate media (as determined by the otherwise unused request *Aura*). Correspondingly, the local *Artefact* master notifies the local *Group* object when it ceases to be a member of the region and the *Group* then pages out the region if no other local artefacts are members of it. Note that only local *master Artefacts* cause the paging of an opaque region - remote master *Artefacts* may have local *proxies* but these do not have local request auras and so should not cause the region to be paged in.

As with aura paging, paging for opaque regions is more conservative (i.e. causes more replication) than would be strictly required by the spatial model. Specifically, no account is taken of the fact that a member of a bidirectionally opaque region can have no possible awareness of non-members. In principle, if all local master artefacts were within a closed region and could not see (hear, etc.) out then the external artefacts need not be paged in. However this is not possible in MASSIVE-2 because the region is itself an artefact which is a member (for replication) of a parent region. This parent region must always be paged in so that the local process will know if the region moves or changes. Also, with overlapping regions there may be artefacts which are members of the region for the purposes of awareness but not replication and which might be incorrectly paged out in this situation. Consequently MASSIVE-2 only attempts to limit paging of the insides of regions, not the outsides. However external awareness driven regions (below) will still page correctly based on awareness adaptation in either direction and might be used to achieve this effect.

Awareness Driven Paging

The third and final form of region paging implemented in MASSIVE-2 is based directly on awareness, rather than aura or region membership. Both the third party object model and MASSIVE-2 allow adaptation effects to be combined with secondary sourcing to allow a region to act as a composite level of detail mechanism or group abstraction: at low awareness values only the region is visible (audible, etc.) while at high awareness values the region's individual members are directly visible. With abstracting regions secondary sourcing provides an overview of regional activity from a distance. For the purposes of replication, when an artefact's awareness of this type of region is below the specified threshold value then there can be no direct awareness of the region's members and so the region can be paged out. When the artefact's awareness of the region passes the threshold value then the region must be paged in so that the region's members can be observed directly.

As with the other forms of paging it is the local *Group* object associated with the region that handles awareness-driven paging. The *Group* object establishes which media are subject to awareness driven paging. For these media it monitors request auras but it does not page in the region when one is found (as would be the case with aura paging). Rather, the *Group* object begins to periodically sample that *Artefact's* awareness of the region in the appropriate medium. Currently this is done every two seconds. When the *Artefact's* awareness of the region exceeds the threshold value then the *Group* object pages in the appropriate media for the region. Correspondingly, when the *Artefact's* awareness falls below the threshold value or the *Artefact* moves out of aura range the *Group* pages out those media (assuming that no other local *Artefacts* remain sufficiently aware of it).

Figure 10.2 on page 148 shows the same virtual world as in figure 10.1 on page 143 but from the perspective of participant *c*. With the aura shown *c*'s process pages in the world group, *W*, both cells, *C1* and *C2*, and the building *B*. The closed room, R, is not paged in because of membership paging while the crowd, X, is not paged in because of awareness paging. Consequently artefacts *a*, *b*, *g* and *k* are locally replicated. However *c* will still have zero awareness of *g* and *k* which are outside its aura and it may not be aware of *a* if the building suppresses outgoing awareness. If *c*'s aura were smaller then cell *C2* would not be paged in (aura based paging). Alternatively, if *c* were to focus more strongly on the crowd, *X* (and the building allowed it to) then *X* might be paged in. Figure 10.2 (a) on page 148 illustrates this spatially, while Figure 10.2 (b) on page 148 shows the flow of information via the region-related multicast groups. Note that *c* may still have some awareness of the contents of *R* and *X* though the secondary source flows from *R* to *B* and from *X* to *C2*, respectively.

Fig.10.2: example replication management for user *c*

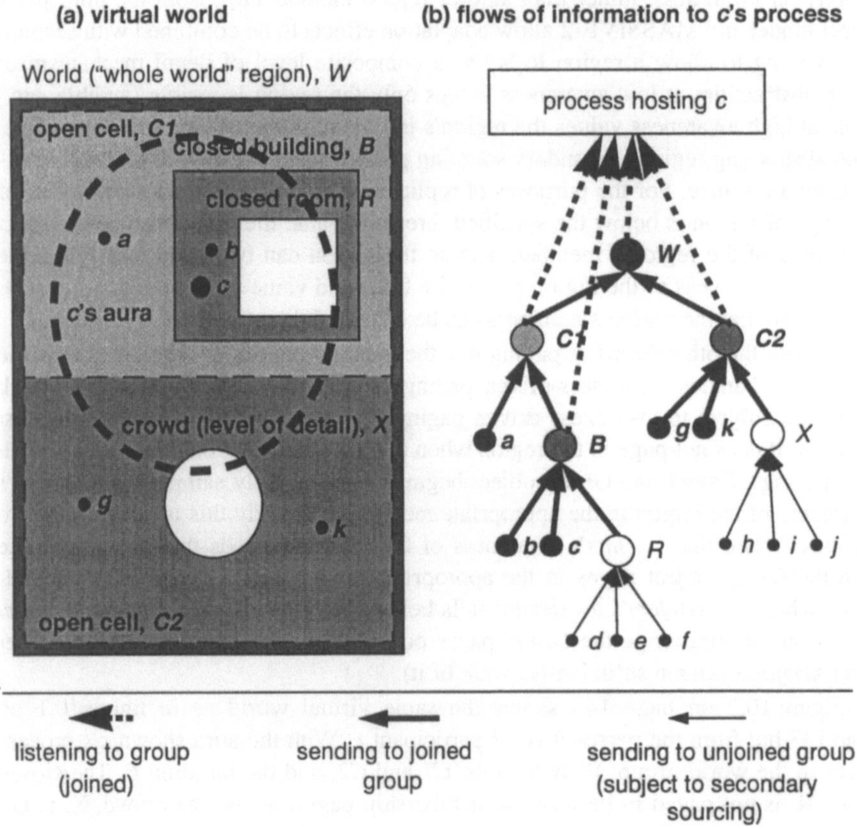

Replication Hierarchies

As shown above replication regions naturally form dynamic spatial hierarchies in MASSIVE-2. Any request aura causes the top-level "region" (the world region) to be paged in. The top level *Artefacts* which are replicated as a consequence will include the next level of regions. When the local proxies for these *Artefacts* are created they will also create local *Group* objects which will manage the paging of these sub-regions based on the different strategies described above and according to the awareness effects of the region. When one of these regions is paged in it may reveal new regions within it, creating new *Group* objects to manage the paging of these regions, and so on. Thus the region hierarchy will unfold in each process as it is needed. This scheme is both flexible and able to accommodate dynamic regions. However there is a risk that paging of and interaction with the top level regions may become a bottleneck. At the cost of some generality and mobility it would be possible to fix the top level group structure and allow paging to begin at the first or second level of the hierarchy (with different participants starting in different regions).

This concludes the description of the implementation of the extended spatial model in MASSIVE-2. It can be seen that the spatial model of interaction in conjunction with third party objects provides a flexible basis for partial replication and multicast group management for CVEs. The rest of this chapter reflects on the use and effectiveness of this approach compared to other strategies for scoping interaction.

10.2. Use

As has already been noted there has been relatively little opportunity to use and formally test MASSIVE-2 to date, particularly with large numbers of users. As such the evaluation in the next section (like that in chapter 6) focuses on modelling and analysing the expected behaviour of this and related systems. The majority of applications and trials (see section 9.2) have not made use of region-related multicasting, but have used the extended spatial model over a fully replicated (per medium) world. As such those applications show that the artefact replication protocol and use of multicasting for audio and updates are correct, but they do not test the replication management aspects of the system, described above. This section describes a demonstration which illustrates the operation of the various forms of replication management described in the previous section.

This demonstration is called "the new audio gallery" (c.f. the "audio gallery" in MASSIVE-1 which was a demonstration piece for awareness-controlled audio). The overall world layout is shown in figure 10.3 on page 150. The world comprises a long series of regions of different types. All of the regions in this demonstration contain objects which are real-time audio sources. The first four regions contain two audio sources each; the last region contains four audio sources. Table 10.1 on page 150 lists the effects of each region. For example, the first region is an open region (which pages on aura collision) and which requires no secondary sourcing. The fourth and fifth regions are awareness-paged and to demonstrate this there is a small (non-multicast) region just in front of each which cuts off audio awareness while a participant is within it.

Figure 10.4 on page 151 shows received unicast and multicast network traffic for a user's client processes as they traverse the full length of the world, passing through each region in turn. Speed of movement is constant and, except for the secondary source regions, all of the regions are the same size. The incoming multicast bandwidth in this example is all audio traffic (the objects are otherwise passive). The incoming unicast traffic comprises region-management traffic and artefact state transfers. It can be seen that the unicast traffic correlates with the user entering and leaving different regions and the paging of regions. Each audio source generates approximately 10Kbytes/s. Note that the total bandwidth inside the regions with secondary sourcing is higher because the observer still receives (but does not play) the second-

Fig.10.3: the new audio gallery world

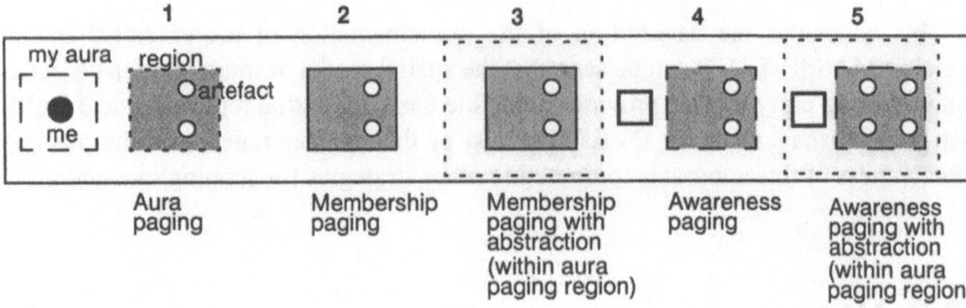

Table 10.1: awareness effects for regions in the new audio gallery

Region	Effect for external observer	Paging type	Secondary source?
1	open	aura	no
2	closed	membership	no
3	closed	membership	yes
4	awareness driven	awareness	no
5	awareness driven	awareness	yes

ary sourced audio. The periodic multicast packets which can be seen between regions are heartbeat messages used to detect process failures.

It is clear from the graph that membership paging is more *specific* than aura paging, i.e. that the region is paged in under more limited and constrained circumstances which match the (user awareness level) third-party characteristics of the region. It is also clear that secondary sourcing provides a different and distinct channel for awareness at a lower bandwidth (assuming, for audio, that there is no compression and that there are two or more sources in the region). The awareness paged examples also show that paging can be driven directly by awareness (the extra closed region has no effect on region membership or aura size, only on audio awareness).

There is one important usability consideration which is worth mentioning here since the evaluation in this chapter focuses on the network rather than the user. It is clear from use to date that the speed at which multicast region paging can be performed is an important consideration. As described in section 8.4 when a local *Group* object decides to page in a medium it has to create local proxies for all of the *Artefacts* in that group. This requires at least one exchange of messages and related processing. For regions with large numbers of members this can introduce a significant delay (several seconds even on a LAN). This delay is significant (and visible) in the case of membership-based and especially awareness-driven region paging. For

Fig.10.4: received multicast and unicast traffic touring the new audio gallery

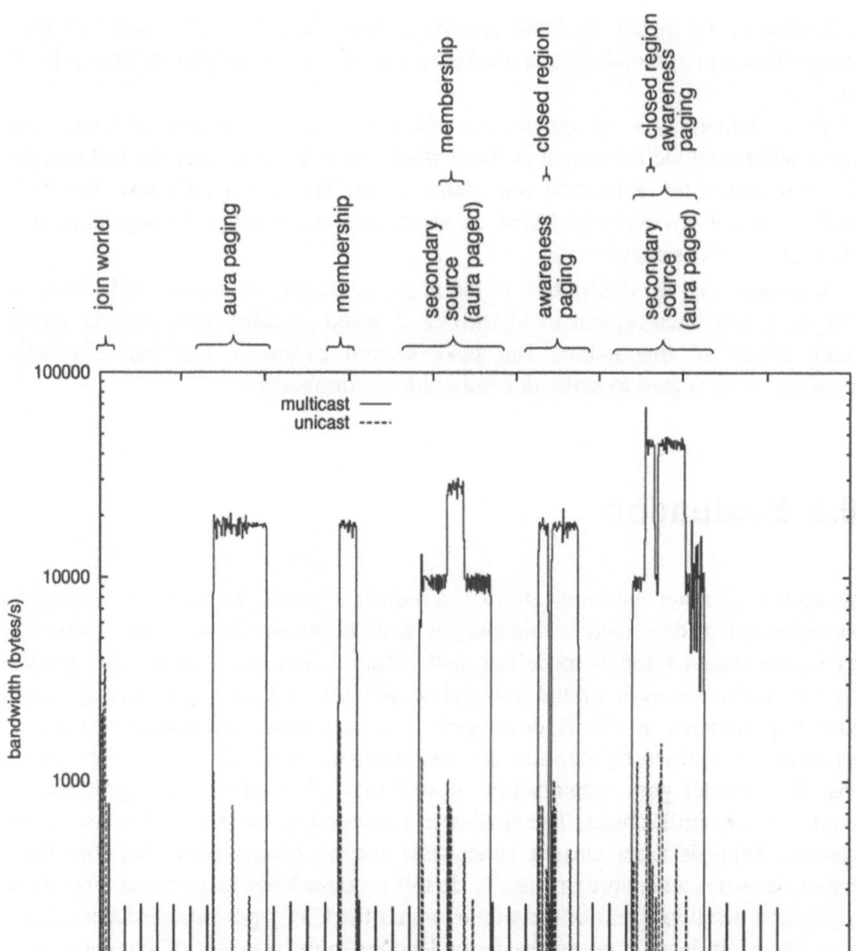

example, on entering a closed room it initially appears to be empty and then its contents appear. Similarly, on becoming sufficiently aware of an level-of-detail region the aggregate view disappears (is suppressed by the awareness negotiation process) but the contents of the region only appear after a delay.

This points to the possible use of look-ahead or prediction and closer coordination of level-of-detail regions with replication. For example closed regions might be paged in initially based on *proximity* rather than membership, in anticipation of the user entering the region. Awareness, including third party effects, would still be calculated normally and would ensure that the objects and information presented to the user are still correct. Ideally this would make use of previous knowledge of user behaviour including speed and direction of movement to make appropriate choices.

Additionally, the group level of detail mechanism in awareness negotiation could delay the transition from low to high awareness with respect to the user, while allowing regions to be paged in. More generally, there is a need to make transitions between direct and secondary sourced views smooth and consistent as viewed by the user.

The implementation and refinement of these techniques is an area for future work. Ideally, additional facilities such as these would hide from the user the fact that partial rather than total replication was taking place. The user would just "inhabit" a seamless virtual environment based on awareness and third party objects as dealt with in the previous chapter.

This concludes the description of the implementation and use of replication and multicast group management in MASSIVE-2 based on third party objects and the spatial model of interaction. The next section evaluates this and alternative approaches with regard to network bandwidth requirements.

10.3. Evaluation

This section explores the implications and requirements of the types of communication management described in this chapter. It demonstrates how issues of this kind may be addressed through modelling and extrapolation from more restricted use. This evaluation revolves around a model of network and participant average bandwidth requirements which is developed here and used to evaluate a range of approaches to multicast group use and management in CVEs and similar applications. This model goes significantly beyond that of chapter 6 in expressiveness, though it has a similar basis. The challenge addressed by the model is to use knowledge and analysis from current small-scale use to reason about possible future large-scale issues and requirements. To do this requires knowledge about typical user activity and behaviour and of the way in which the CVE application relates these to the network. It is also necessary to understand how patterns of communication develop and change as more simultaneous users are added to the system. The components of this model are shown pictorially in figure 10.5 on page 153. The key components of the model are introduced below.

Task and application requirements are considered explicitly in this model whereas tele-conferencing was the only application considered for MASSIVE-1. This component of the model represents the way in which the system will be used and the kinds of worlds and world content which may be expected. This is an area in which this model extends significantly that of chapter 6.

As in chapter 6 the model of user behaviour employed here is based on measured user activity and behaviour in the ITW trials with MASSIVE-1. See sections 6.3.1 and 6.3.2, appendix A and [Greenhalgh et al., 1997] for fuller details of the data and analysis which underlies this component.

Fig.10.5: a predictive model of CVE network behaviour

Model components	Answers the question...
Task/application requirements	What do people want or try to do?
User behaviour	What particular actions do people do and when?
Process behaviour	How does the application respond?
Distribution architecture	What communicates with what?
Communication Protocols	How is information exchanged?
Network communication	What happens "on the wire"?

Again, as in chapter 6, the model components quantifying process behaviour and communication protocols have been obtained by detailed analysis of small-scale applications, relating individual key user and application actions to messages and other significant events.

In chapter 6 the distribution architecture was for MASSIVE-1, i.e. based on unicast peer-to-peer communication managed by spatial trading. However this section generalises this aspect of the model to include network architectures which make effective use of multicast communication. This is another area in which this model extends the previous one.

The final outputs from the model are predictions of network load in terms of total average network bandwidth and also the average bandwidth received by a single participant's machine. The final network traffic model is used to explore:

- the relative impact of state size vs. update or streamed bandwidth;
- the difference between explicit state transfer and the use of heartbeat messages;
- the benefits of multicasting compared to unicast;
- the effect of different choices of replication unit;
- the effect of different forms of replication management;
- the influence of secondary sourcing and abstraction.

Finally, the MASSIVE-1 and MASSIVE-2 systems are represented in the model to illustrate its use and to indicate the relative scalability of these two systems with respect to network bandwidth requirements.

The specification and modelling of task and application requirements is in some ways the hardest and most significant part of the whole model. This is the area which is dealt with first.

Note that all of the parameters defined in the model are summarised in table 10.6 on page 178.

10.3.1. Application Model

The "application" referred to here is the overall purpose or task for which the system is being used (rather than the individual processes which make up the system). This will determine what kinds of interaction must be supported, what types and quantities of information must be exchanged and may also have a strong influence on actual user behaviour (the next element of the model). The application model defines the relationships between a number of key conceptual components of a CVE session. These components are: worlds, regions, artefacts, participants and scopes of interaction. The nature and number of these components are parameters of the application model which allow a range of tasks and applications to be described. The concepts, overall model and specific parameters are described in turn.

Concepts

The application model is made up of a number of components: worlds, regions, artefacts, participants and scopes of interest. These are defined in turn below.

- Worlds. All interaction is assumed to occur within a world. Each world is a disjoint space, with its own coordinate system. Most CVEs have some notion of worlds which may also be linked via portals or other mechanisms as in MASSIVE-1 and DIVE [Hagsand, 1996].
- Regions. As described in chapter 2 many multi-user virtual reality systems (especially those which employ multicasting) divide each world into smaller cells or zones which play a role in scoping or facilitating communication. Examples include regional third party objects in MASSIVE-2 and hexagonal tiles in NPSNET [Macedonia et al., 1995].
- Artefacts. These are the tangible (visible, audible, etc.) objects which populate the virtual environment. Most or all interaction in the CVE is mediated by artefacts. Examples include the representations (embodiments or avatars) of users, scenery, buildings, useful objects, elements of a visualisation, vehicles, etc.
- Participants. These are the agencies or active processes within the virtual environment and include human user client processes as well as software agents and applications. A participant is typically represented within the virtual environment by one (or more) artefacts. Participant processes also observe surrounding activity in the virtual environment, e.g. to present this information to a human user, or to control an application.
- Scopes of interest. The interests of each participant within the virtual environment are described by a scope of interest. In the original spatial model this was the participant's various medium-specific auras. With third party objects this should also take account of third party effects as considered in section 10.1. The

same idea is implicit or explicit in most of the systems considered in chapter 2. The participant wishes to know about artefacts within this scope, but need not know about artefacts outside it.

The communication task of a CVE runtime system is to provide all current participants with appropriate and timely information about all artefacts which are within their scope of interest. This information may include graphical geometry, position and orientation, articulations and gestures, audio, video and other forms of interaction and communication. This is in the context of potentially mobile artefacts and normally mobile participants. This is illustrated in figure 10.6 on page 155.

Fig.10.6: main components of the CVE application model

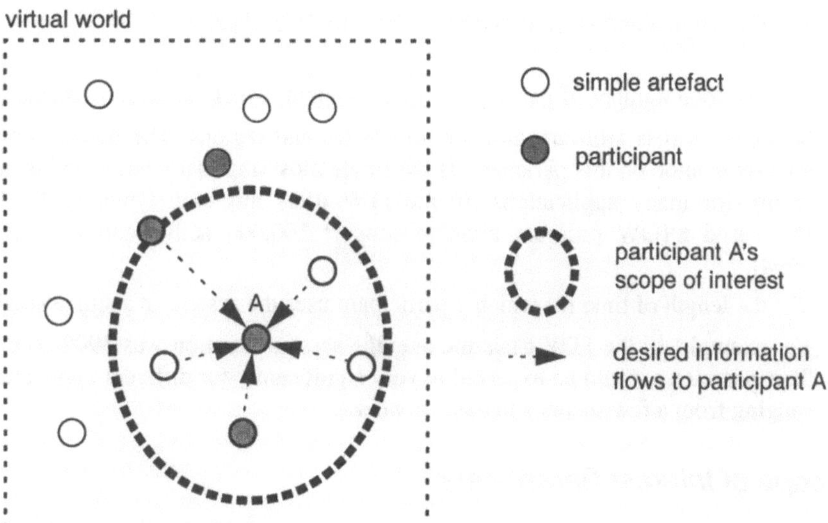

The following sections formalise the characteristics of artefacts, participants and scopes of interest within the model. These define the model's key parameters. Consideration of worlds and regions is deferred until the distribution architecture is considered since they are intimately related.

Artefact Parameters

In this simple model there is one application/task-level parameter concerned with artefacts which is defined below.

- N_A: the total number of non-participant artefacts in the entire system (spread across arbitrary numbers of worlds and regions). One of the key dimensions of scalability is the dependency of communication on this parameter. In this model non-participant artefacts are assumed to be passive, i.e. static and generating negligible numbers of updates. Typical values for N_A would be from 1 (the simplest possible non-empty world) to many millions. For example, in the ITW trials it was typically around 10 while for many visualisation applications it would be several hundreds or thousands. Note that all of the parameters defined in the model are summarised in table 10.6 on page 178.

Participant Parameters

Each participant is represented in the system by an active artefact. For simplicity, these participant embodiments are not included in the passive artefact count, N_A, above. If a system or application includes agents or other active artefacts then they should be counted as participants (or participant-equivalents) for the purposes of this model. The parameters relevant to participants are defined below.

- N_P: the total number of participants in the system, which, as with artefacts, may be spread across arbitrary numbers of worlds and regions. The dependency of communication on this parameter is the single most important dimension of scalability for many applications. In the ITW trials this way typically 10. The DIS-based STOW program aims to support 100,000 active entities [ARPA, 1994].
- T_s: the length of time for which a participant uses the system in a single session. For example, in the ITW trials the average session duration was 3900 seconds. This parameter would be expected to vary significantly for different applications, ranging from a few seconds to days or weeks.

Scope of Interest Parameters

Each participant has their own scope of interest. In this model it is characterised by the following parameters.

- I_A: the number of passive artefacts which fall within an average scope of interest. This model assumes a generally consistent density of artefacts. Typical values range from 1 (a minimal meeting space as in some of the ITW meetings) to 10^6 or more in large visualisation applications.
- I_P: the number of participants which fall within an average scope of interest. Typical values range from 0 (for a single user application!) to tens of thousands (a virtual wembley stadium). Example values are 10 for the ITW meetings and 800 suggested in [Macedonia et al., 1995]. If participants are assumed to be evenly distributed (like artefacts) then I_P and I_A are related by the expression:

$$\frac{I_A}{N_A} = \frac{I_P}{N_P} \qquad \text{(Equation 10-1)}$$

- M: mobility of interest, i.e. the number of times during a session that the participant moves (e.g. by changing world or moving within a world) such that the artefacts and participants within their scope of interest are completely replaced by new ones. This must be at least 1 which corresponds to their arrival at the beginning of the session. T_S/M is therefore the average time over which the scope of interest changes once. In the ITW trials the average value for M was 13, corresponding to 12 inter-world transitions during an average meeting and an average of 300 seconds (5 minutes) spent in each world. For the military simulation application analysed for NPSNET area of interest management in [Macedonia et al., 1995] such a change should not occur more than once an hour, i.e. $M \leq T_s/3600$.

So, to summarise the model so far, there are N_A artefacts and N_P participants within the system. The artefacts are static and passive while the participants and mobile and active (generating ongoing information such as updates, audio, etc.). Each participant stays in the system for T_s seconds on average. Each of these participants is interested at any one time in I_A other artefacts and I_P other participants and their interests change M times during the session as a result of their own movement.

10.3.2. User Model

The previous section has sketched out the overall task-oriented structure of collaborative virtual environments and has defined a number of application-dependent and scalability-related parameters. This section further defines the nature and activities of participants within this environment, and is based on observations and analysis of the ITW trials (see Appendix A). This is very close to the user model using in section 6.3.1 except that a number of parameters and assumptions have now been moved into the application domain (above). The key elements of the model are listed below.

- All users have 3D graphical and real-time audio interaction capabilities.
- The most significant user events are moving and speaking.
- Users move 20% of the time (see appendix A, section A.2.1). However limiting the update rate (decoupling it from frame-rate) means that movement occurs in more update intervals, so that users effectively move (with respect to generated traffic) about 30% of the time with 1 second updates.
- Users speak (or rather send network audio data) 25% of the time (appendix A, section A.3.1).
- Communication and interaction are assumed to be independent of frame rate.

These values influence the values of process behaviour parameters, below.

10.3.3. Process Behaviour and Communication Protocols

These two areas are considered together because the focus of this model is on network traffic rather than internal process behaviour and this is determined by the combination of process behaviour (what the process tries to do) and communication protocols (what kinds and sizes of messages are involved in communicating this). As in the application/task component of the model there are a number of parameters here which may be varied to reflect differences between tasks and also differences between systems such as the efficiency and flexibility of artefact representation. These parameters are defined below.

- S: the size (in bytes) of the state of an average artefact (or participant embodiment), i.e. a snap-shot description of an artefact. This may include (pointers to) geometry, audio and video information, identity, text attributes and awareness information (e.g. focus and nimbus parameters). In DIS [IEEE, 1993] this corresponds to the size of an Entity State Protocol Data Unit (ESPDU) which is 152 bytes including packet headers, however in the normal DIS approaches which use unreliable communications throughout and periodic retransmission of state this will effectively be zero with regard to this model since state snap-shots are not generated on request. In MASSIVE-2 with geometry specified by name (comparable to use of URLs rather than in-place geometry) this is 1628 bytes for a typical artefact. In MASSIVE-1 with geometry specified in full this is about 13 Kbytes (see table 6.2 on page 84).
- B_P: the bandwidth (in bytes per second) generated by a participant. In DIS with 3 updates per second this would be 456 bytes/second. From analysis and observation of MASSIVE-2 in use a typical value would be 2495 bytes/second, comprising audio 25% of the time (2387 bytes/second, over 95% of the total), movement 30% of the time, limited to 1 Hz updates (43 bytes/second), gestures and graphical "mouth" (53 bytes/second) and keep-alive messages (12 bytes/second); this makes use of the user model values from the previous section.
- B_A: the bandwidth (in bytes per second) generated by a passive artefact. In MASSIVE-2 and other systems with an explicit reliable state transfer phase this will be zero. In the DIS approaches which use unreliable communications throughout and periodic retransmission this will be non-zero. For a message sent every 5 seconds this gives 30.4 bytes per second. For the exponential heartbeat timer back-off scheme of [Holbrook et al., 1995] this might reduce to, say, every 120 seconds giving 1.3 bytes per second average with a correspondingly longer delay to discover artefacts in a newly joined world or region.

The use or otherwise of state transfer (e.g. when joining worlds and regions) is considered in the next section which deals with this and other aspects of distribution architecture.

10.3.4. Distribution Architecture

This key component of the model deals with the structuring of communication and includes the effects of task or application parameters such as the number of participants. This section also deals with worlds and regions as the external manifestations of communication and replication management. Possible distribution architectures are introduced below in terms of a number of largely orthogonal choices concerning: replication; state transfers; and use of multicasting. The final model allows these alternatives to be compared more systematically with regard to network and participant bandwidth requirements.

Replication

Perhaps the most important consideration is how communication is structured to reflect each participant's individual scope of interest. This has two components: first, the units in which replication and communication are handled, and second, the way in which replication and communication are managed. There are a number of alternative choices for the unit of replication or communication. The specific choices are described and analysed using this network traffic model in sections 10.3.11 through 10.3.13. There are also a range of replication management schemes which might be adopted and these are described and analysed in section 10.3.14.

Independently of the specific approach adopted, the effects of the choice of replication unit and the form of replication management are reflected in this model by the two key parameters defined below.

- A_I: the *accuracy* of replication compared to a participant's ideal scope of interest.

  Accuracy is the most significant parameter for describing and comparing approaches to awareness management in CVEs. The accuracy of a system may be defined in the terms of figure 1.3 on page 12 as the amount of *presented* information at the user interface divided by the amount *requested* and/or *achieved* at the network level. In terms of artefacts this will be the number of artefacts presented to the user divided by the number of artefacts which the user's machine receives information about over the network. This will depend on both the unit of replication and on the form of replication management. The subsequent parts of the analysis indicate values for this parameter for different approaches.
- E_M: the "efficiency" with which mobility is handled. Specifically, in the world and region-based replication schemes an artefact (i.e. a participant's embodiment) will need to announce its arrival and state each time it enters a new world or region for the purposes of replication. In this model each participant is assumed to change its area of interest M times in a session and so 100% efficiency will be defined as only being required to announce its state M times during the session (this is independent of newly arriving *observers* obtaining the initial state of an

artefact). As defined, values of E_M can vary from 0 (continuous state transfers) to ∞ (if these announcements never occur).

State Transfer

The subject of state transfers (i.e. whether to have them or not) has already been raised in section 10.3.3 in the context of DIS heartbeat messages. This consideration is reflected in the choice of values for S and B_A as described above. There are also timeliness considerations when replicating new units which may be particularly important with smaller replication units and with more flexible forms of replication management such as that used in MASSIVE-2.

Another issue concerns the effect of changing scope and then returning: is a new state transfer required, or can some form of caching be used? To use caching there must normally be some mechanism to verify that the cached data is up to date, and this mechanism must be significantly more light-weight that a new state transfer. This might be reflected in this model by a lower effective value for artefact state size, S. However this is not considered directly in this model.

Multicasting

The final architectural/implementation consideration is the use of multicast communication (or otherwise). The change from unicast-based communication to multicast was one of the main motivations for creating a second version of MASSIVE. This was based on simple analyses such as that given in section 6.3.6. In this model of CVE requirements this choice will be reflected directly in the derived expressions for total network bandwidth (i.e. one for total unicast and one for total multicast).

10.3.5. Overheads

Before giving final expressions for network and participant bandwidth requirements it is necessary to consider management overheads such as spatial trading or multicast group joins and leaves. Different replication schemes also require different numbers of units of replication. In turn, each unit of replication will have corresponding resource requirements. In particular, where multicasting is being used each unit of replication will require one or more multicast groups (e.g. it might require one for each medium). These issues are reflected in the parameters below.

- N_R: the total number of replication units used by the system.
- N_{RI}: the average number of replication units of interest to a single participant at any one time. This will depend on the choice of unit of replication; this is dealt with in section 10.3.13. Because of the inaccuracy with which scope of interest is

represented a participants machine will actually replicate on average N_{RI}/A_I units to achieve this.

Paging in or out of a replication group will have costs in terms of multicast group management (if multicasting is being used), changing resource reservations (if used) and may also require external support to trigger it (such as MASSIVE-1's spatial trading service). For the purposes of the network model two additional parameters are defined below.

- S_{RI}: the network traffic (bytes) associated with each paging of a replication group (this is the total for paging in and out). This includes application messages which may be required to cause paging as well as lower-level multicast group management messages. A participant's scope of interest changes M/T_S times per second and therefore the average rate at which units of replication will be paged in and out by a single participant will be $(MN_I)/(T_SA_I)$ (see the definition of N_{RI}, above).
- B_{RI}: the continuous bandwidth (bytes per second) associated with replication management generated by each participant. For example, in MASSIVE-1 this involves keeping the spatial trader up to date with aura information as the participant moves. This does not include actual aura collision notifications which are a function of the replication scheme employed and are represented by S_{RI}, above.

The world and region based replication schemes require that artefacts change replication groups as they move. This results in additional state transfers (already included in the bandwidth expressions) and also multicast group management and resource reservations changes (if used).

10.3.6. Putting It Together

This section summarises and draws together the components of the model and leads on to the derived expressions for network and participant bandwidth requirements.

- Referring back to the application model (section 10.3.1 and figure 10.6 on page 155) there are N_P simultaneous participants exploring and working in one or more virtual worlds which contain a total of N_A passive artefacts.
- Each participant has an ideal scope of interest which describes their information requirements; this includes I_A artefacts and I_P other participants at any time and changes because of their own movement M times during each session (which lasts T_S seconds).

- The participant's ideal scope of interest is represented in the distribution system with an accuracy of A_I.
- As the participant moves about they have to announce their presence and state on M/E_M occasions during each session. This state (like the state of any artefact) comprises S bytes of information as seen by the network and will be received by I_P/A_I other participants (assuming information flows between participants are approximately symmetrical).
- Each participant also generates updates and streamed data at an average rate of B_P bytes per second. This is received by other participants and may comprise audio data and other update messages and is suitable for multicast dissemination (some of it reliably).
- To manage replication each participant also generates B_{RI} bytes per second of management traffic.
- The distribution system might represent the participant's ideal scope of interest by N_{RI} replication units (depending on the replication approach adopted). Because of inaccuracies and compromises in representing scopes of interest each participant's machine actually replicates N_{RI}/A_I units.
- This means that the distribution system replicates I_A/A_I artefacts and I_P/A_I participants at any moment. It also receives a corresponding $(I_A/A_I)B_A$ bytes per second of artefact updates (which is zero in many systems) and $(I_P/A_I)B_P$ bytes per second of other participant updates and streamed data.
- The distribution system's view of the participants scope of interest will change (as does the ideal scope of interest) M times in each T_S second session. Each unit of replication will cause S_{RI} bytes of management traffic each time it is paged in and out.
- Also, the distribution system must obtain $(I_A + I_P) S/A_I$ bytes of state information for each change in a participant's scope of interest (in addition to the state messages sent out by participants as a result of their own movement, above).

Different applications and tasks will be reflected in different values for N_A, N_P, I_A, I_P, M and T_S. Different systems, representations of artefacts and use of media will be reflected in the values for S, B_A and B_P; these also reflect the choice between explicit state transfers and periodic heartbeat messages. Different approaches to replication and communication organisation and management will result in different values for A_I, E_M, N_{RI}, B_{RI} and S_{RI}. The use of multicasting or otherwise will determine whether total network bandwidth is impacted by copying packets to multiple participant observers.

10.3.7. Network Traffic Model

If multicast communication is assumed for mobile participant state announcements and for artefact and participant updates then the total network bandwidth will be:

$$\frac{SMN_P}{T_S E_M} + \frac{S(I_A + I_P)MN_P}{A_I T_S} + B_A N_A + B_P N_P + \left(\frac{MN_{RI} S_{RI}}{A_I T_S} + B_{RI}\right) N_P$$

(Equation 10-2)

where the five terms correspond to: multicast announcements/state transfers of mobile participants; unicast state transfers due to changes in participants' scopes of interest; multicast heartbeat messages (if any) from static artefacts; updates and streamed data from participants; and replication unit overheads. Note that this ignores any additional overheads associated with multicast communication (e.g. to ensure reliable delivery).

The traffic above is the network total and subsets of it will be received by each participant's own machine. Specifically, each participant's machine will be observing (and handling the information in) an average bandwidth of:

$$\frac{SI_P M}{T_S E_M} + \frac{S(I_A + I_P)M}{A_I T_S} + \frac{B_A I_A}{A_I} + \frac{B_P I_P}{A_I} + \frac{MN_{RI} S_{RI}}{A_I T_S} + B_{RI}$$ (Equation 10-3)

If unicast communication is used throughout the system rather than a mixture of unicast and multicast as above then the total network bandwidth (c.f. equation 2) will be:

$$\frac{SI_P MN_P}{T_S A_I E_M} + \frac{S(I_A + I_P)MN_P}{T_S} + \frac{B_A I_A N_P}{A_I} + \frac{B_P I_P N_P}{A_I} + \left(\frac{MN_{RI} S_{RI}}{A_I T_S} + B_{RI}\right) N_P$$

(Equation 10-4)

With unicast communication the participant bandwidth will be unchanged.

The following sections use this network traffic model to explore:

- the relative impact of state size vs. update/streamed data bandwidth;
- the difference between explicit state transfers and heartbeat messages;
- the benefits of multicasting compared to unicasting;
- the effect of different choices of replication unit (in terms of accuracy and overheads);
- the effect of different forms of replication management;
- the influence of secondary sourcing and abstraction.

Finally the model is applied to the MASSIVE-1 and MASSIVE-2 systems.

10.3.8. State and Updates

In the approaches considered here initial state transfers (to create artefact proxies) and subsequent updates and streamed data such as audio have different and distinct communication requirements. In particular updates can be readily multicast whereas at least half of state transfers must be unicast on demand. Furthermore, an application may have many more artefacts than participants, with a correspondingly higher demand for state transfers (which are needed for both artefacts and participants) than for updates (which are needed for participants alone).

For simplicity this section assumes ideal scopes of interest and movement (i.e. A_I and E_M equal to one) and that overheads are negligible (i.e. ignoring the contribution of B_{RI} and S_{RI}). Then, ignoring heartbeat messages (which are considered in the next section), the total bandwidth when using multicast (equation 2) simplifies to:

$$\left((I_A + I_P + 1)\frac{SM}{T_S} + B_P\right)N_P \qquad \text{(Equation 10-5)}$$

and the total bandwidth with only unicast (equation 4) simplifies to:

$$\left((I_A + 2I_P)\frac{SM}{T_S} + I_P B_P\right)N_P \qquad \text{(Equation 10-6)}$$

Note in each case that there is a single state-related term and a single update-related term. Note also that I_A and hence the number of artefacts contributes to the state-related term but not to the update-related term. Table 10.2 on page 164 shows illustrative values for these terms (per participant) for a number of different scenarios using figures from MASSIVE-2 ($S = 1628$ bytes, $B_P = 2495$ bytes/s).

Table 10.2: state and update bandwidth for a range of application scenarios

Scenario	I_A	I_P	$\frac{T_s}{M}$ seconds	Multicast total, bytes/s (part.)	Unicast total, bytes/s (part.)
teleconference	10	10	1200	29 + 2495 = 2524	41 + 24950 = 24991
visualization	1000	10	600	2743 + 2495 = 5238	2768 + 24950 = 27718
race!	100	10	5	36142 + 2495 = 38637	39072 + 24950 = 64022

The bandwidth for multicast and unicast are shown in total and for state and updates separately. For multicast, the first scenario (tele-conferencing) is dominated by updates, the second (collaborative data visualisation) is fairly balanced while the third (a high-speed race, perhaps a game) is dominated by initial state transfers. However, in the unicast case the updates are more significant than in the multicast case because of the additional I_P fan-out factor, so that even the third scenario is only just balancing out state transfers and update traffic.

With regard to a single participant, the single participant received bandwidth (equation 3) simplifies to:

$$(I_A + 2I_P)\frac{SM}{T_S} + I_P B_P \qquad \text{(Equation 10-7)}$$

and exhibits the same balance between state and updates as for unicast total network bandwidth, above.

These values are for long-term average bandwidth. It will also be important to consider the burstiness of each. For example, state transfers will be concentrated around world transitions whereas updates will be relatively constant. On the other hand it may be easier to enforce flow control on large state transfers than on many independent sources of updates and audio streams.

10.3.9. Heartbeat Messages

The previous section shows that in some situations explicit state transfer bandwidth could dominate update and streamed data, especially for multicast-based systems. This section considers the alternative approach to state transfers of using periodic multicast heartbeat messages to convey state. Making the same simplifying assumptions as above but using heartbeat messages the total bandwidth when using multicast (equation 2) simplifies to:

$$\left(\frac{N_A}{N_P}B_A + B_P\right)N_P \qquad \text{(Equation 10-8)}$$

Similarly, the single participant bandwidth (equation 3) simplifies to:

$$I_A B_A + I_P B_P \qquad \text{(Equation 10-9)}$$

These are directly comparable to equations 5 and 7 for explicit state transfers. If state is resent every T_H seconds (and assuming B_P already includes sufficient state information) then B_A will be S/T_H. In each case the update bandwidth is the same whereas the state requirement differs. Table 10.3 on page 166 compares the state requirements directly.

Table 10.3: comparison of explicit state transfer vs. heartbeat approach

Transfer type	Total state-related bandwidth	Participant state-related bandwidth
Explicit state transfer	$(I_A + I_P + 1)\frac{SM}{T_S}N_P$	$(I_A + 2I_P)\frac{SM}{T_S}$
Heartbeat	$\frac{N_A S}{T_H}$	$\frac{I_A S}{T_H}$
Ratio (explicit/ heartbeat)	$(I_A + I_P + 1)\frac{N_P}{N_A} \cdot \frac{MT_H}{T_S}$	$\left(1 + \frac{2I_P}{I_A}\right)\frac{MT_H}{T_S}$
Ratio (for uniform distribution): $\frac{I_A}{N_A} = \frac{I_P}{N_P}$	$\left(1 + \left(1 + \frac{1}{I_p}\right)\frac{N_P}{N_A}\right) \cdot I_P \cdot \frac{MT_H}{T_S}$	$\left(1 + 2\frac{N_P}{N_A}\right) \cdot \frac{MT_H}{T_S}$

For any normal application it is easy to ensure that $N_P \leq N_A$, i.e. that there are at least as many artefacts as participants (more extreme application would be rather degenerate in terms of content). With this constraint the first term in each of the final ratios is limited to the range 1.0 to 3.0. The key term in balancing explicit state transfer requirements against the heartbeat approach is MT_H/T_S, which is the relative frequency of state transfers due to mobility compared to heartbeat messages. So if heartbeat messages are less frequent than changes of world or area of interest then the heartbeat approach will yield a reduction in bandwidth to an individual participant. However this is self-defeating, because the participant is relying on these heartbeat messages to learn what artefacts are in its new scope of interest. So heartbeat messages must occur reasonably soon, and certainly significantly more often than changes of interest. Consequently for useful behaviour this factor will always be significantly less than 1.0.

With regard to total network bandwidth (rather than single participant bandwidth, considered above) the issue is slightly less clear because of the additional I_P factor for explicit state transfers, which reflects the fact that multicasting is not being used in the explicit state transfer case and so I_P unicast messages are needed compared to each multicast message. Consequently if I_P is large, total network bandwidth is the limiting factor and state transfers dominate updates (see previous section) then the heartbeat approach would be appropriate, resulting in a reduction in total bandwidth requirements at the cost of increased participant bandwidth.

However, a more general approach would be to allow explicit multicasting to be used when a significant number of participants attempt to replicate a region or world (or whatever) at the same time. This might use an extra "initial state" multicast group associated with each unit of replication. This would require reliable multicasting and highly dynamic multicast groups, but this is an area of rapid development at the present time.

10.3.10. Multicast and Unicast

The first observation concerning the choice between multicast and unicast communication is that, for the same replication approach, each participant receives the same bandwidth (equation 3). So this choice does not affect the amount of information which a participant machine receives and has to deal with. It does however affect the total network bandwidth and the number of packets which must be *sent* by a single machine.

Comparing equations 2 (multicast) and 4 (unicast) the second term (unicast state transfer) is unchanged but the first and fourth terms (multicast state transfer and participant updates, respectively) are increased by a factor of I_P/A_I, i.e. the average number of participants who are within interest range of another participant, allowing for the efficiency of the system in representing their scopes of interest. The third term (heartbeat messages) changes by a factor of $(I_A N_P)/(A_I N_A)$ however this term is unlikely to be relevant in a unicast-based system since the use of heartbeat messages for state transfer is designed to *avoid* the need for either unicast communication or for reliable multicast communication.

So, in general, the unicast state transfer component is unaffected but the state transfers due to self-movement and most significantly the participant update bandwidth increase by a factor of I_P/A_I. This factor reflects the use being made of multicasting and places an *upper* limit on the potential cost of using unicast compared to multicast (when updates dominate state transfers). Referring to the prior section on state versus updates it is apparent that, at least for MASSIVE-2, update bandwidth is quite likely to dominate or at least be comparable in demands with state transfer and therefore the use of multicasting will yield appreciable reductions in total network bandwidth requirements. For example, table 10.2 on page 164 shows reductions for MASSIVE-2 of 90%, 81% and 40% for a teleconferencing scenario, a visualization scenario and a game scenario, respectively. If total bandwidth remains the limiting factor then some form of multicast state transfer should be investigated (see the previous section).

10.3.11. Unit of Replication

The choice of replication unit and form of replication management is reflected in the network traffic model by its influence on the A_I and E_M terms and in the management overheads and system requirement as represented by S_{RI}, B_{RI}, N_R and N_{RI}. The four principle alternative units of replication are: universe, world, region and artefact. These are introduced below.

- Universe. The original SIMNET system used network broadcasting [Johnston, 1987] and consequently imposed no structure on communication. Effectively (whether it is presented to users in this way or not) every participant shares a single virtual world without regions or other structuring and the unit of replication is just "everything".
- World. A natural enhancement to the above approach is to structure the virtual "universe" into a number of disjoint virtual worlds with communication and replication performed for each world individually. DIVE version 2 [Carlsson and Hagsand, 1993] used this approach as, effectively, did NPSNET prior to the introduction of cells [Macedonia et al. 1994].
- Region. Replication and communication management may also be organised based on regions or an equivalent concept. This is necessarily more limited in scope than a single world, but may include a number of artefacts. Section 10.1 has described how this approach is used in MASSIVE-2. The same kind of approach is found (without awareness-based management) in other systems such as Spline [Barrus et al., 1996] and Broll's work [Broll, 1997]. Use of multicast groups associated with artefact hierarchies in DIVE 3 [Hagsand, 1996] can also be considered in this category.
- Artefact. Finally, replication and communication may be performed for individual artefacts. This was the case in MASSIVE-1 for example (see chapter 6).

As noted above the choice of replication unit dictates (together with management of replication): the accuracy of replication; and the associated overheads. These are considered in turn.

10.3.12. Accuracy of Replication

This section considers the potential accuracy of the different choices of replication unit listed above for the case of a simple open world, i.e. without closed regions or abstractions. More complicated situations are considered in the subsequent section dealing with replication management. Table 10.4 on page 169 shows values for A_I and E_M for each of the choices of replication unit described above. Reflections for each case are included below.

- Universe. The efficiency of interest for the universal replication case is just the fraction of the total number of artefacts or participants which actually fall within a scope of interest (however it is defined). The universal replication case avoids the need for all but the first state transfer when a participant joins the system and so E_M is greater than 1 (as defined). In fact the network traffic model begins to break

Table 10.4: accuracy of interest and efficiency of movement in different replication schemes

Unit of replication	A_I	E_M	Parameters	Constraints
Universe	$\frac{I_A}{N_A}$	M		
World	$\frac{I_A N_W}{N_A}$	$M \geq E_M \geq 1$	N_W, the number of worlds	$I_P \leq \frac{N_P}{N_W}, I_A \leq \frac{N_A}{N_W}$
Region (open square 2D cells)	$\frac{\pi N_{RI}}{4(\sqrt{N_{RI}}+1)^2}$	$\frac{1}{\sqrt{N_{RI}}}$	$N_{RI} = \frac{I_A N_R}{N_A}$	$\frac{I_A}{N_A} = \frac{I_P}{N_P}$
Artefact	1	$\frac{1}{I_P}$ [a]		

a. This efficiency is with respect to total network traffic and reflects the loss of potential multicasting. As far as a single observer is concerned it will be 1.

down for universal replication because mobility is essential meaningless with respect to replication in this case. However since this approach is inherently unscalable this shortcoming will not be addressed here.

- World. The efficiency of interest for the multiple worlds case depends entirely on how closely the members of a world match a participant's scope of interest. The table assumes that participants are spread evenly between worlds and introduces the parameter N_W, the number of active worlds. Clearly, the number of artefacts or participants in a world must be at least the number in the scope of interest if it is to be satisfied. Also the efficiency of movement will depend on whether movement within worlds is significant. With a few large worlds this tends towards the single world case with inaccurate interest but super-efficient movement. At the other extreme, with many minimally sized worlds the accuracy of interest is good but the efficiency of movement falls to one (the participant can only change their scope of interest by jumping to a new world).
- Region. The values for regions in this table are based on the use of square 2D regions of equal size with no form of occlusion or other awareness limitation, with a circular scope of interest (of any size) and with a uniform 2D spatial distribution of artefacts and participants. For comparison, the NPSNET hexagonal tile approach in [Macedonia et al., 1995] has an example with 19 regions considered at a time. For a circular scope of interest the worst case radius to fit within these 19 regions is $\sqrt{7}L$ where L is the length of a cell's side. Hence in that approach

$$A_I = \frac{\pi(\sqrt{7}L)^2}{19\left(\frac{3\sqrt{3}}{2}L^2\right)} = \frac{14\pi}{57\sqrt{3}} = 0.445$$

For square cells and a similar circular scope of interest ($N_I = 8.45$) the average value of A_I using the expression in the table is 0.435 (the effect of using squares rather than hexagons is offset by assuming that regions are replicated only as required, rather than always replicating the same number of regions).

- For single artefacts it is assumed that the spatial trading service (or its equivalent) can deal exactly with scopes of interest so that artefacts are replicated only as and when required. There is no indirect communication as in the previous cases (e.g. via a region or a world); each artefact is "just itself". Consequently the artefact sends its state as a result of individual aura collisions (or the equivalent) with individual participants. The table assumes that all I_P of the participant's neighbouring participants change for each change of interest scope and therefore I_P individual state requests are generated and satisfied.

General conclusions, including consideration of overheads and management, are given at the end of section 10.3.14.

10.3.13. Replication Overheads

In these various replication schemes different number of units of replication will be required. In turn, each unit of replication will have corresponding resource requirements. In particular, where multicasting is being used each unit will require one or more multicast groups (e.g. one for each medium). In table 10.5 on page 171 columns 2 and 3 show the total number of replication units in the system, N_R, and the number of interest to a single participant (subject to representational inaccuracies), N_{RI}/A_I, for the different schemes considered. For example in the world-based scheme there is one replication unit for each world, i.e. N_W in total, and each participant's machine deals with exactly one unit (one world) at a time. The rate at which replication units are paged in and out are shown in columns 4 and 5 (in total and per participant). The rate at which participants move between replication units (as a sender) is shown in columns 6 and 7. So, for example, in the world-based case each participant stays in the system for T_S seconds and moves (changes worlds) M times in that period so that they page in new worlds (i.e. new replication units) at an average rate of M/T_S per second (column 5). There are N_P participants in the total system giving rise to the corresponding value in column 4. With per-world replication each participant sends to the same group which they listen to and they move (as a sender) to a new group at the same rate as they change worlds, i.e. M/T_S times per second (column 7). Again there are N_P participants in total giving the value in column 6.

Table 10.5: replication-related overheads, per participant and in total

Unit of replication	Units (total) N_R	Units (part.) N_{RI}/A_I	Pages/s (total)	Pages/s (part.)	Moves /s (total)	Moves /s (part.)
Universe	1	1	$\frac{N_P}{T_S}$	$\frac{1}{T_S}$	$\frac{N_P}{T_S}$	$\frac{1}{T_S}$
World	N_W	1	$\frac{MN_P}{T_S}$	$\frac{M}{T_S}$	$\frac{MN_P}{T_S}$	$\frac{M}{T_S}$
Region	N_R	$\frac{N_{RI}}{A_I}$ [a]	$\frac{MN_{RI}N_P}{T_SA_I}$	$\frac{MN_{RI}}{T_SA_I}$	$\frac{MN_P}{T_SE_M}$	$\frac{M}{T_SE_M}$
Artefact	N_A+N_P or N_P [b]	$\frac{I_A+I_P}{A_I}$ or I_P/A_I	$\frac{M(I_A+I_P)N_P}{T_SA_I}$	$\frac{M(I_A+I_P)}{T_SA_I}$	$\frac{N_P}{T_S}$	$\frac{1}{T_S}$

a. see table 10.4 on page 169 for the definition of N_{RI} in this case.

b. for state transfer (rather than heartbeat-based) systems in which artefacts can *never* change the replication-related requirements for artefacts may be avoided.

General conclusions are given at the end of the next section.

10.3.14. Replication Management

As well as the choice of unit of replication another issue of replication management concerns the flexibility of the replication management process with reference to the user's ideal scope of interest (however that may be specified). For example, replication management in MASSIVE-2 can take account of opaque boundaries and awareness-driven abstractions which are not taken into account in MASSIVE-1 or in the simple Area of Interest management proposed for NPSNET [Macedonia et al., 1995]. This may have a large impact on the accuracy with which the system represents a participant's scope of interest in different situations. For example, the approach of NPSNET is good for open terrain but potentially highly inaccurate in densely occluded environments, where the line-of-sight approach of RING [Funkhouser, 1995] would be ideal. However this would depend on using individual artefacts or very many small regions as the units of replication.

Because MASSIVE-2 uses an explicit computational model of awareness it can accurately represent and reason about a participant's scope of interest. Consequently it is able to model a range of situations with relative accuracy. This is one of the key benefits of the spatial model of interaction with third party objects. One estimate of scoping accuracy is developed below.

In this traffic model a participant's ideal scope of interest is represented in the first instance by ideal values for I_A and I_P which are assumed to be derived in some way from a more abstract specification of interaction scope (e.g. using the spatial model or based on typical use). In open terrain this scope of interaction is (in principle) largely related to what is possible within the constraints of the medium, for example how far you can see on a clear day. A particular value of I_A or I_P would therefore imply a certain density of artefacts or participants, say $D_{A\min}$ and $D_{P\min}$, respectively. In a more enclosed space architectural features such as walls and rooms are the principle constraints on the scope of interaction. This implies a correspondingly higher density of artefacts or participants within this kind of environment, say $D_{A\max}$ and $D_{P\max}$, respectively. Using an open-style scope in a closed-style environment (without taking account of its closure) would therefore include many more artefacts or participants than required, specifically:

$$A_I = \frac{D_{A\min}}{D_{A\max}} = \frac{D_{P\min}}{D_{P\max}}$$

For example, if an ideal scope of interest was a 1 km radius circle in open terrain or a 6 metre square room then A_I could be as small as 0.001%!

Summary

With consideration to management, overhead and accuracy it is possible to make a number of observations and recommendations concerning replication. These are given below for each size of unit considered here.

- Universe. With the availability of multicast communication there is no reason to adopt this simplistic approach to replication.
- World. Replication based on worlds is the simplest viable approach. It requires the fewest replication units and therefore the fewest multicast groups. It also has less need (or opportunity) for replication management. However the accuracy of this approach depends entirely on matching interaction and interest groups exactly to distinct worlds. This approach also means that there is no continuity for awareness changes: a participant must jump to a completely new world.
- Region. The region approach is an intermediate between the world and artefact-based approaches. It requires an intermediate number of multicast groups (some small multiple of N_R) and has a correspondingly intermediate management cost. However a region-based approach can provide much better continuity

of awareness and movement than a strictly world-based approach. It can also yield much better efficiency or accuracy in a range of situations and applications. This is demonstrated in MASSIVE-2 which employs the spatial model of interaction and third party objects to manage region-based replication.

- Artefact. This approach has the largest resource requirements (at least one multicast group per artefact) and the finest-grained management requirement (such as distributed spatial trading in MASSIVE-1). However it can also guarantee excellent accuracy in any kind of application (assuming that the ideal scope of interest has a computable basis).

The "ideal" approach might be a flexible combination of world, region and artefact based replication. Worlds provide a common low-complexity base-line which can be supplemented with regions as appropriate. Finally, artefact based replication may be appropriate for unusually demanding artefacts or media, such as real-time video. In this case the cost of unnecessary replication and communication would be large and the use of another multicast group would be easy to justify.

10.3.15. Abstraction and Secondary Sourcing

Abstraction and secondary sourcing are not dealt with directly in the model presented above but it is possible to reason about their effects.

A key motivation for using abstraction and secondary sourcing is to reduce the load on individual participant's machines by replacing awareness of many individual artefacts with a few abstractions. Adding secondary sources may increase total network multicast bandwidth requirements since the secondary sourced data is in addition to the normal directly available information flows. For abstractions the additional data is based on the data which is already available with the express intention of producing a simpler, more compact and less demanding partial representation of it. So the maximum total bandwidth requirements for abstractions will normally be a small fraction of the total bandwidth. This assumes that the abstraction process is effective (but if it wasn't then hopefully it would not be used).

Consider the case in which each participant will be satisfied with partial (abstracted) awareness of a fraction f_I of its ideal scope of interest and that viewing an artefact through an abstraction requires a fraction f_B of the equivalent resources for direct viewing (e.g. if an abstraction represents $1/f_B$ artefacts using the same state and bandwidth as one artefact). Then for the simplified form of the network model (equations 5 and 7), ignoring accuracy, efficiency and overheads, the total multicast bandwidth would be:

$$\left(((1-f_I(1-f_B))(I_A+I_P)+(1+f_B))\frac{SM}{T_S}+(1+f_B)B_P\right)N_P \quad \text{(Equation 10-10)}$$

Note that multicast bandwidth increases by a factor of $1 + f_B$ reflecting the additional secondary sourced information but unicast state transfers fall by a factor of $(1 - f_I(1 - f_B))$ reflecting the replacement of state transfers for multiple individual artefacts with the equivalent abstractions. The bandwidth for a single participant would be:

$$(1 - f_I(1 - f_B))\left((I_A + 2I_P)\frac{SM}{T_S} + I_P B_P \right) \qquad \text{(Equation 10-11)}$$

showing the improvement $(1 - f_I(1 - f_B))$ throughout. For example, if abstractions were sufficient for 75% of artefacts, and abstraction resulted in a 10-fold reduction in resource requirements on average (i.e. $f_I = 0.75$ and $f_B = 0.1$) then the single participant bandwidth would be reduced by 67% while the total network bandwidth would, for the three scenarios in table 10.2 on page 164, change by +9.1%, -30.5% and -61.8%, respectively (for the teleconference, visualisation and race). Larger values of f_I and smaller values of f_B would produce larger savings.

It is apparent that secondary sourcing and abstraction have the potential to significantly reduce the load on a single participant if abstracted views of many artefacts are acceptable to the user. Whether this is the case will depend on the application. In many scenarios the use of abstractions may also significantly reduce the *total* bandwidth requirements because of the reduced demand for state transfers.

10.3.16. Examples

This section concludes this analysis by applying the network traffic model to the MASSIVE-1 and MASSIVE-2 systems described in this thesis. These are considered in turn.

MASSIVE-1

MASSIVE-1 replicates based on artefacts and uses explicit state transfers rather than DIS-style heartbeat messages. The following system-specific parameters apply for MASSIVE-1 based on the analysis in chapter 6: S = 13.2 Kbytes, $B_A = 0$, $B_P = 4.6$ Kbytes/s, S_{RI} = 2100 bytes and B_{RI} = 1400 bytes/s (approximately, including inter-client communication). Using artefact-based replication (and assuming the sufficiency of spatial trading to represent scope of interest) $E_M = 1$, $A_I = 1$ and $N_{RI} = I_A + I_P$. Networking is unicast and so total network bandwidth (from equation 4) is:

$$\left(15400 \cdot (I_A + 2I_P)\frac{M}{T_S} + 4600 \cdot I_P + 1400 \right) N_P \qquad \text{(Equation 10-12)}$$

Similarly the per-participant bandwidth (from equation 3) is:

$$15400 \cdot (I_A + 2I_P)\frac{M}{T_S} + 4600 \cdot I_P + 1400 \qquad \text{(Equation 10-13)}$$

Chapter 6 also neglected I_A and took M/T_S to be $1/60$ per second, giving a total network bandwidth of $(4850I_P + 1400)N_P$ bytes/s (c.f. equation 1 on page 85, $B = N(4800M + 1400)$, where $M = I_P$ and $N = N_P$). For this case the single participant bandwidth is $4850I_P + 1400$ bytes/s.

MASSIVE-2

MASSIVE-2 performs replication based on worlds and regions, uses explicit state transfers like MASSIVE-1 and uses multicast communication for updates. The following system-specific parameters apply for MASSIVE-2 (see section 10.3.3): S = 1628 bytes, $B_A = 0$, $B_P = 2495$ bytes/s, $S_{RI} = 1624$ bytes and $B_{RI} = 0$. The total network bandwidth will be:

$$\left(\left(1628\left(I_A + I_P + \frac{1}{E_M}\right) + 1624N_I\right)\frac{M}{T_S A_I} + 2495\right)N_P \qquad \text{(Equation 10-14)}$$

The single participant bandwidth will be:

$$\left(1628\left(I_A + I_P + \frac{I_P}{E_M}\right) + 1624N_I\right)\frac{M}{T_S A_I} + \frac{2495I_P}{A_I} \qquad \text{(Equation 10-15)}$$

For square cells as analysed in section 10.3.12 (and table 10.4 on page 169):

$$A_I = \frac{\pi N_I}{4(\sqrt{N_I} + 1)^2} \qquad , E_M = \frac{1}{\sqrt{N_I}}$$

For example, consider $M/T_S = 1/60$, $I_A = 0$ (as for MASSIVE-1, above) and $N_I = 8.45$, $A_I = 0.435$ and $E_M = 0.344$ (comparable to NPSNET cells as in section 10.3.12). This gives a total network bandwidth of $(125I_P + 3100)N_P$ bytes/s. This is a 38-fold reduction compared to MASSIVE-1 for large values of I_P. For I_p = 10 (as in the ITW trials) the reduction in bandwidth requirements is by a factor of 11. The single participant bandwidth is $5877I_P$. So the reduced accuracy of open cells results in a typically higher average bandwidth for each participant than in MASSIVE-1 in this type of environment.

This reduced accuracy and correspondingly increased per-participant bandwidth is a general consequence of using a larger unit of replication (regions rather than artefacts). This choice is in turn motivated by the adoption of multicast communication, with system and network restrictions and overheads linked to the number of multicast groups employed.

However when an environment includes closed regions then MASSIVE-2's exploitation of third party object effects would make it more accurate than MASSIVE-1's aura-only approach (and other approaches which assume a nominally open space). This would cause a corresponding reduction in participant bandwidth.

10.4. Summary and Conclusions

This chapter has described how the spatial model of interaction and third party objects are used in MASSIVE-2 to manage artefact replication and multicast groups.

By using an explicit computational model of awareness with third party effects the system is able to represent a range of situations from open terrain to closed rooms with reasonable accuracy. This is achieved through three forms of replication management: aura-based, membership-based and awareness-based. All three are demonstrated in the new audio gallery world (described in section 10.2).

The majority of this chapter has developed a network bandwidth model for CVEs based on experience gained over the course of the work presented in this thesis. This traffic model is used to analyse a number of distribution and communication issues in CVEs and similar systems. The results are summarised below.

- State and updates. The model shows that the relative requirements of state transfers compared to (multicastable) updates vary widely for different application scenarios. In extreme applications (such as the "race" example in table 10.2 on page 164) state transfers can dominate, especially when multicasting is used for updates. In situations such as this multicast state transfers or group aggregates might be used to reduce the total network bandwidth requirements (though multicast state transfers will not affect the participant bandwidth).
- Heartbeat-based state transfers. For applications which involve many mutually aware participants (i.e. large I_P) and which are limited by total network bandwidth rather than participant bandwidth then using heartbeat messages to perform state transfer can reduce total network bandwidth requirements. However this will typically result in a significant increase in single participant bandwidth, as well as the need to trade off timeliness of information against bandwidth requirements through the choice of T_H, the heartbeat time. In general explicit state transfers are more appropriate. More specific (reliable) multicast state transfers or abstractions could be used if state bandwidth requirements remained the limiting factor for scalability.
- Multicast and unicast. The benefits of using multicasting in terms of total network bandwidth can be very great, especially if the update bandwidth is relatively high

(e.g. with video) or if some form of multicasting is also to be employed for state transfers. The use of multicasting reduces the number of messages which a participant has to send but it does not reduce the number which they will receive. In fact the use of multicasting may lead to the use of a *less accurate* form of replication management (see below) so that each participant actually has to deal with more information. However, except on very high-speed LANs (e.g. workgroup ATM) total network bandwidth limitations will mean that multicasting will be a necessary choice for significant numbers of mutually aware participants.

- Replication unit. Replication can be performed at a number of levels of granularity, classified here as "universe", "world", "region" and "artefact". The universal replication approach is unnecessarily limited. The world-based approach is simple and useful, but depends on being able to organise interest and interaction to match. The region approach is more flexible than the world approach as demonstrated in MASSIVE-2. Finally the artefact approach is potentially the most accurate (a participant only replicates what they need) but the management overheads will typically be greatest as will the network resource requirements (e.g. the number of multicast groups or reservations required). The best compromise might be to combine general region-based replication with artefact-based replication for extremely demanding artefacts (e.g. which include real-time video streams).
- Replication management. Depending on the choice of replication unit replication management can be performed in a number of ways. The goal is to replicate only those artefacts which are of direct interest to a given participant. This requires a flexible and expressive way of representing the participant's interest and an accurate way of mapping this onto units of replication. This is one strength of MASSIVE-2 and the spatial model of interaction with third party objects. More limited forms of replication management (e.g. using open cells or based solely on occlusion) will be either less accurate or more limited in the range of applications which they can address.
- Secondary sourcing and abstraction. Secondary sourcing and abstraction have the potential to significantly reduce the load on a single participant if abstracted views of many artefacts are acceptable to the user; this will depend on the application. Also, where unicast state transfers are the dominant component of total network bandwidth then the use of abstractions may also significantly reduce total bandwidth requirements.

The model and methodology used in this analysis could also be applied and generalised to other situations and to other classes of application. The notion of awareness (which need not be explicit in the application) provides the essential basis for the model and allows reasoning about, for example, accuracy, which is one of the key components and outcomes of the model.

This concludes part II of this thesis which has considered MASSIVE-2 and the third party object concept. The next and final chapter draws together the various themes running through this thesis and presents the final conclusions and suggestions for future work.

Table 10.6: summary of network model parameters

Parameter	Meaning
A_I	The *accuracy* of replication compared to a participant's ideal scope of interest.
B_P	The bandwidth (in bytes per second) generated by a participant.
B_A	The bandwidth (in bytes per second) generated by a passive artefact (often zero).
B_{RI}	The continuous bandwidth (bytes per second) associated with replication management generated by each participant.
E_M	The "efficiency" with which mobility is handled.
I_A	The number of passive artefacts which fall within an average scope of interest.
I_P	The number of participants which fall within an average scope of interest.
M	Mobility of interest, i.e. the number of times during a session that the participant moves such that the artefacts and participants within their scope of interest are completely replaced by new ones.
N_A	Number of artefacts (passive) spread over all worlds.
N_P	Number of simultaneous participants spread over all worlds.
N_R	The total number of replication units used by the system.
N_{RI}	The average number of replication units of interest to a single participant at any one time (ideally).
N_W	The total number of worlds over which artefacts and participants are distributed
S	The size (in bytes) of the state of an average artefact or participant's embodiment.
S_{RI}	The network traffic (bytes) associated with each paging of a replication group (in and out, total).
T_s	The length of time for which a participant uses the system in a single session.

11. Conclusions

This final chapter begins in section 11.1 by summarising the work presented in this thesis chapter by chapter. Section 11.2 then presents the main conclusions and contributions of this work organised in terms of collaborative virtual environments: their philosophy, theory, realisation, use and modelling. Section 11.3 reflects on the work presented here. Section 11.3.1 identifies areas for future work. Section 11.3.2 gives more personal reflections on the work that has been done and the way in which it has been done. Finally, section 11.3.3 concludes this thesis by reflecting on how this work relates to the "big picture" of computer science in society.

11.1. Summary

This thesis is concerned with the design and realisation of large-scale collaborative virtual environments, that is, virtual environments which actively support collaboration between large numbers of simultaneous users. The central theme around which this work has been organised is that of "awareness". This rather general term reflects both the sociology of interpersonal communication and the flow and transfer of information in a computational context. Returning to figure 3 of the introduction (which is reproduced as figure 11.1 on page 180) CVEs can be viewed in terms of a "cycle of awareness" which connects the dispersed users via their local computing resources to "the rest of the world" (i.e. the network). To realise this cycle of awareness it is necessary that awareness have some concrete and computational representation within the system. For the work presented here this is provided by the spatial model of interaction (which is reviewed in chapter 3). This computational model of awareness influences, manages and in some cases articulates interaction between the user and the virtual environment as presented to them by their local computing resources. This same model of awareness must also control and manage interaction and communication over the network. In particular it is necessary that such an awareness model be mapped to a feasible and efficient network communication architecture.

Existing multi-user virtual reality systems can be used to build collaborative virtual environments. However they do not explicitly consider awareness as a tool for

Fig.11.1: the cycle of awareness

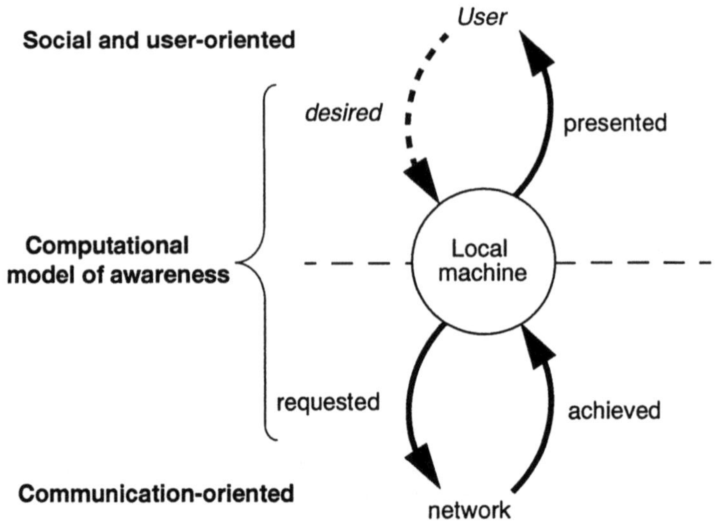

collaboration. Instead the awareness controlling facilities which are present are solely motivated by the desire to *restrict* awareness in some situations in order to enhance system performance and/or scalability. Thus, rather than a full model of awareness they have one or more mechanisms and metaphors for scoping interaction, i.e. for limiting the extent over which interaction must be considered (in whatever dimension). These various approaches to scoping interaction are surveyed in chapter 2. From this a number of issues are brought out which may be used to classify this type of scoping and to guide system designers.

The work presented in this thesis has passed through two complete phases of implementation and evaluation which are described in this thesis in chapters 3 through 6 and 7 through 10, respectively. The two prototype systems constructed are MASSIVE-1 and MASSIVE-2. Both are multi-user virtual reality systems and may justifiably be regarded as *collaborative* virtual environments because each is based on a comprehensive and ubiquitous computational model of awareness.

In the case of MASSIVE-1 the awareness model used was the pre-existing spatial model of interaction of Benford and Fahlén (as reported in [Benford and Fahlén, 1993]). The spatial model of interaction was reviewed in chapter 3 as background to the description of MASSIVE-1 given in chapter 4. Chapters 5 and 6 are the heart of part I of this thesis and each presents details of the system's implementation, of its use and its evaluation.

Chapter 5 focuses on the MASSIVE-1 system and the spatial model as they are viewed by a typical user (i.e. a participant in a collaborative environment). This chapter describes the realisation of the awareness relationship by which awareness is negotiated between both users and objects. This is based on direct negotiation over individual peer-to-peer connections using unicast network communication. The evaluation component of chapter 5 considers awareness in use based on observation, direct experience and informal interviews and discussion with system users. This draws on MASSIVE-1's use for over 20 wide-area meetings and many other sessions. The system (and the model) meets its goals in a number of respects, however a significant number of shortcomings and unresolved issues are apparent. The system has supported collaboration and a number of interesting and suggestive events have been observed including "natural" uses of space and unanticipated interaction. However problems remain with the spatial model as implemented and with the system itself. In general there is a need to improve the ease of navigation and the degree to which a participant is aware of their surroundings. In respect of the spatial model it is apparent that awareness, focus and nimbus (the key components of the model) need to be made more directly visible and controllable within the interface. Further work is also needed to refine and "fine tune" the expression and calculation of focus, nimbus and awareness. The final shortcoming apparent in the spatial model is the extremely limited support for contextual factors in interaction (e.g. being in a room compared to being in an open park).

Chapter 6 provides the network-oriented complement to chapter 5 and deals with the network architecture and communication requirements of MASSIVE-1. This chapter describes the implementation of the aura relationship which is the basis of spatial scoping of interaction in this system. Aura management is formalised in the context of distributed systems as "spatial trading". This is a development of more traditional attribute-based trading such as that of ODP [ITU-T, 1995] combined with the virtual reality technique of collision detection. Spatial trading is distinguished by the persistence and explicit representation of both offers and requests and the support for efficient incremental modifications to offers and requests. The evaluation of chapter 6 analyses the total network bandwidth requirements of MASSIVE-1 as a function of the total number of simultaneous participants and the size of interaction groups which are formed. This analysis shows that spatial trading is potentially much more efficient than dealing directly with *all* of the potential interactions in a virtual world. This is directly comparable to Area of Interest management in NPSNET [Macedonia et al., 1995]. The same analysis shows that, compared to using multicast-based communications, the unicast-based approach of MASSIVE-1 typically requires an order of magnitude more network bandwidth.

Part II of this thesis focuses on two of the issues identified in part I. From chapter 5 it addresses the lack of support for context in awareness negotiation while from chapter 6 it addresses the omission of multicast communication from the original implementation. The principle theoretical contribution of this thesis is in chapter 7 which presents the concept of "third party objects" as a proposed extension to the spatial model of interaction. Third party objects are defined in terms of the effects which they have on awareness and in terms of the way in which they are activated. The model defines two potential effects: adaptation, which modifies existing awareness relationships, and secondary sourcing, which introduces new indirect forms of awareness. Activation is defined in terms of the various awareness relationships which exist between a third party object and the other objects on the world (which it affects). Three particular patterns of activation are highlighted: membership, common focus and a hybrid case. It is proposed that third party objects might be used to realise a range of awareness effects including buildings, rooms, open cells, crowds, floor control, cross-medium adaptation, anonymity and remote awareness. The second prototype system, MASSIVE-2, is introduced in chapter 8. MASSIVE-2 implements a large subset of the third party object model in addition to the original spatial model of interaction. It does this using a complementary multicast-based network architecture. Chapters 9 and 10 parallel chapters 5 and 6 in part I and each presents details of the implementation of MASSIVE-2, of its use and its evaluation.

Chapter 9 describes the implementation of awareness negotiation in MASSIVE-2, paying particular attention to the realisation and capabilities of spatial third party objects (i.e. regions). This is illustrated by a number of applications and demonstrations including the Arena, the Panopticon Plaza and a collaborative WWW browser (WWW-3D) which has been extended to incorporate regions or "data districts". As in chapter 5, the evaluation of chapter 9 considers the use of the system and the model. The third party object concept is shown to be useful in realising a range of effects including rooms and buildings, crowds and also more unusual forms of awareness context such as unidirectional boundaries and anonymity services. However a number of shortcomings remain which correspond closely to those found for the MASSIVE-1 system in chapter 5. In particular third party objects exacerbate the need to make awareness and its components more directly visible and accessible to users.

Chapter 10 describes how the spatial model of interaction and third party objects are used in MASSIVE-2 to scope and manage multicast-based network communication. There are three forms of interaction scoping which are exploited in MASSIVE-2: aura range; membership of closed regions (e.g. buildings and rooms with opaque boundaries); and awareness-driven level-of-detail groups abstractions. Each of these results in more efficient use of network and computational resources compared to other simpler approaches (e.g. based on disjoint worlds). Chapter 10 evaluates MASSIVE-2 and the spatial model plus third party objects through the development and analysis of a predictive model of network bandwidth requirements. This model of network requirements is significantly more general than that of chapter 6. This model is used to consider and compare various approaches to interaction scoping and communication management in CVEs and similar systems. In particular approaches based on worlds, regions (as in MASSIVE-2) and artefacts are compared and contrasted. In general, a region-based approach appears to offer the best compromise between accuracy and flexibility on the one hand and overheads and network resource requirements on the other. This network model also shows (in qualitative terms) that more sophisticated awareness management schemes such as that used in MASSIVE-2 can be significantly more accurate and efficient than the simpler and less flexible schemes used in contemporary systems. Finally, the model demonstrates that the use of network multicasting significantly reduces total network bandwidth requirements. However this may lead to the choice of a less accurate approach to scoping interaction and correspondingly higher demands placed on individual participant's machines.

11.2. Main Contributions

This section draws out the main original contributions of this thesis to a range of areas in and around Computer Science. As noted at the beginning of the previous section the central interests of this thesis revolve around the notion of collaborative virtual environments and in particular the way in which a concept and model of user awareness can be used to structure and facilitate communication and collaboration. Readers interested the areas of Computer Supported Cooperative Work (CSCW) and Computer-Human Interaction (CHI) (and also sociology and psychology) may wish to focus on the user-oriented aspects of awareness, as it is made available to and directly affects the user. Readers interested in the areas of distributed systems and multimedia may wish to focus rather on the way in which awareness is realised and used to structure and manage communication within computer-based systems, for example viewed as a form of session management or quality of service control. Readers interested in computer networking may wish to concentrate on the way in which the concept of and support for awareness shapes the system's overall communication requirements and brings the various and variable requirements of end-users into the sharp relief. Finally, the general reader, while gaining an overview of all of the these aspects, may wish to consider the philosophy of collaborative virtual environments and awareness-driven communication as an alternative paradigm for the design and use of computer-based systems.

The following sections identify the key contributions of this work and are structured with reference to a number of different aspects of collaborative virtual environments: their philosophy, theory, realisation, concrete experience and prediction or modelling of general use. These are dealt with in turn.

11.2.1. Philosophy

This work contributes to the philosophy of collaborative virtual environments and computational models of awareness primarily through its provision of the MASSIVE-1 and MASSIVE-2 systems as prototypes of this approach (see chapters 4 and 8). For example, analyses of meetings held in MASSIVE-1 carried out by Bowers, O'Brien and Pycock ([Bowers, O'Brien and Pycock, 1996] and [Bowers, Pycock and O'Brien, 1996]) have yielded new insight into the relationships between real and virtual activities: the transfer of real-world behaviours into the virtual world, and the importance of the ongoing co-existence of "real" and "virtual" activity and interaction. The author has also been actively involved in the exploration of other issues for collaborative virtual environments such as the representation of context in interaction (as presented in this thesis as third party objects) and issues of embodiment and representation of users in CVEs [Benford, Bowers et al., 1997].

11.2.2. Theory

This thesis proposes the third party object concept as an extension to the pre-existing spatial model of interaction of Benford and Fahlén [1993]. The third party object concept is presented in chapter 7 and allows this extended spatial model of interaction to explicitly represent and reason about the effects of context or environment on awareness relationships. Third party objects may represent aspects of interaction context such as rooms and buildings, crowds, common artefacts or more abstract factors such as membership of a group or the control of a chairperson. Third party objects are defined in terms of their effects (what they do) and their activation (when they do it). Two classes of effect are defined: adaptation, which involves modification of existing awareness relationships, e.g. amplification or suppression; and secondary sourcing, which introduces new indirect forms of awareness. The combination of both effects is particularly useful to realise group effects including aggregation and abstraction of the group as a whole (however it may be defined). Third party objects are activated or controlled by other awareness relationships, specifically those between the third party and the objects which it is influencing. Three primary patterns of activation are identified: membership of the third party, representing a group of some kind; common focus on the third party as an object of common interest; and a hybrid form of activation which combines membership and focus on the object.

Key observations of the third party object concept are listed below.

- Third party objects are first class objects within the total spatial model framework, and can themselves exploit focus, nimbus and aura to manage their own interaction and operation.
- Third party objects, because they are also first class objects, can apply to one another, allowing the construction of combined, linked or nested patterns of effects.
- Third party objects can be as dynamic and flexible as any other object. For example, in a CVE they can be dynamically introduced and can be mobile and change in size and effect over time.
- Third party object activation is based on awareness negotiated between the third party and *both* of the other objects in the interaction. This may be contrasted with the original notion of adapters in the spatial model of awareness which was based on one of the objects only. This represents a significant increase in the expressiveness of third party objects compared to simple adapters.
- The combination of adaptation and secondary sourcing (the two kinds of third party effects) within a common framework is necessary to successfully manage the introduction of abstraction and other kinds of secondary sources within the system.
- As noted above, third party objects can be used to represent and realise notions of groups, regions (closed boundaries), common interest and indirect awareness.
- As demonstrated in MASSIVE-2 (see chapter 10) the third party concept can be exploited by a compatible distribution architecture to give more accurate or effi-

cient patterns of information dissemination in an awareness-based system when compared to the original spatial model of interaction and to other current medium-independent approaches.

11.2.3. Realisation

The work presented in this thesis has included two complete prototype collaborative virtual environment systems: MASSIVE-1 and MASSIVE-2. The first is a virtual reality tele-conferencing system based on the original spatial model of interaction and employs spatial trading and unicast-based peer-to-peer awareness negotiation to manage inter-participant communication through 3D graphics, audio and text (see chapter 4). The second is a general-purpose collaborative virtual environment system which includes the third party object concept in addition to the original spatial model facilities (see chapter 8). MASSIVE-2 can be used to support tele-conferencing and also other applications such as collaborative information visualisation. These two prototype systems demonstrate the viability of the presented models as the basis for an operational distributed implementation.

These systems demonstrate concrete realisations of the concepts presented and may inform and inspire future systems. Specifically, MASSIVE-1 prototypes and demonstrates the use of spatial trading which is a potential new generic service for distributed systems which extends the attribute based trading of ODP [ITU-T, 1995] in line with the philosophy of CVEs and in particular the concept of aura as found in the original spatial model of interaction (spatial trading is presented in chapter 6, while chapter 3 reviews the original spatial model of interaction). Spatial trading differs from attribute-based trading (for example in ODP) in a number of respects: both offers and requests persist within the trading space; offers and requests are subject to incremental updating, which is efficiently supported; requests are always comprehensive (requesting all matches rather than a single match); and matching includes a notion of distance. This reflects the philosophical shift from relatively closed and well-defined tasks to more open and ongoing interactions and potential interactions based on the concept of awareness.

MASSIVE-2 employs multicast group and communication management which is directly controlled by the third party objects within the environment. This allows it to accurately map between idealised user requirements for awareness and resulting communication requirements. This exploits provisions of third party objects for representing spatial constraints on interaction. This also allows MASSIVE-2 to share and allocate multicast groups according to appropriate contextual constraints. In turn, this allows interaction scoping to be performed effectively with restricted numbers of multicast groups (compared to the total number of virtual artefacts).

11.2.4. Experience

The MASSIVE-1 and MASSIVE-2 systems are also the basis for gaining user experience of these models of awareness. Both systems have been used for numerous small group laboratory meetings and demonstrations. The MASSIVE-1 system has been used as the primary meeting support tool for the BT/JISC funded Inhabited The Web (ITW) project, which included 17 wide-area group meetings between six sites with up to ten participants in any meeting (see [Greenhalgh et al., 1997] for more details of these meetings). The MASSIVE-2 system has been used to stage the "MASSIVE" real/virtual poetry performance as part of the Nottingham NOWninety6 arts festival. This allowed ten members of the public at a time to view and interact in each of six virtual poetry performances (see section 9.2.1). These and other experiences form the basis of the user-oriented evaluations in chapters 5 and 9. Other researchers have also used these systems as a basis for evaluation and exploration, most notably Bowers, O'Brien and Pycock ([Bowers, O'Brien and Pycock, 1996], [Bowers, Pycock and O'Brien, 1996]). The main observations based on use of the system are listed below under the headings: success; awareness and virtual presence; and third party objects.

Success

- Both systems have been successful in supporting effective and enjoyable collaborations in meetings and less formal settings.
- There have been examples of social browsing and spontaneous collaboration using the systems. This was one of the motivations behind the space and awareness-based approach of the spatial model.
- It is apparent that the audio medium is critical for supporting synchronous communication and collaboration.
- Bowers, O'Brien and Pycock [1996] observe some characteristics of real-world bodily activity in communication being transferred (on occasion) to the virtual world, suggesting that the users can, at least to some extent, make effective use of the available movement and gesture facilities.

Awareness and Virtual "Presence"

- It is possible to create small group situations in which focus and/or nimbus are necessary and effective. However their use in small relatively focused groups does not normally appear to be essential (especially when multiple connected worlds are also provided to scope interaction).
- In the desktop configuration used in these systems the graphical medium limits (eliminates) peripheral awareness of surrounding activities. This is especially apparent in an informal group context when other participants may be anywhere around the user in the virtual space.
- Compared to physically co-located interaction there are frequent breakdowns of communication which appear to be due in part to technical difficulties and in part to the relative lack of non-vocal expression (especially semi-automatic back-channel signals such as nodding and gaze direction).

- There are occasional break-downs of presence and embodiment when participants "leave" their virtual bodies without this being apparent in the virtual world. For example, they may answer the phone or leave the office while still connected to the session.
- Navigation remains awkward, especially for novice users. This is a general issue for virtual reality and 3D visualisation.

Third Party Objects

- MASSIVE-2 demonstrates useful and effective spatial third party effects, in particular closed rooms and buildings, crowds, and a number of other boundary effects (in the Panopticon Plaza for example, see section 9.2.3).
- MASSIVE-2 also demonstrates that third party abstractions can be used to dynamically reduce communication load, particularly in the audio medium.
- It is apparent that awareness and the influence of the third party object must be made more visible to the participants so that they can understand and reason about the world as it is experienced. This is particularly important with the more unusual third party effects such as asymmetric boundaries and medium-specific boundaries.
- Third party objects can be used to effectively manage network communication and multicasting (see for example the New Audio Gallery in section 10.2).
- The temporal characteristics of this communication management are also significant. In the current implementation there are significant transient discontinuities in the presented audio and graphics in these situations, for example in the transition from an abstracted view to a direct view (e.g. seeing "into" a crowd).

Experience of the system has also yielded information about the way in which the system is used and the resulting demands which are placed on the communications infrastructure and user machines. This is considered in the next section.

11.2.5. Modelling of Use

This thesis has presented a general predictive modelling framework for distributed application use and requirements, in particular, for network communication (see section 10.3). This involves analysis and modelling of both user activity and of system and network behaviour. These are considered separately below.

User Activity

Profiles and characterisations of user activity have been created based on auto-generated application and network logs from the ITW trials with MASSIVE-1. This logged information has been used to systematically analyse aspects of user activity and behaviour within the system (see appendix A). For the meetings in question these give sample values for: time spent moving, correlation of movement between users, speed of movement, occurrence of group transitions between worlds, occurrence of return visits to worlds, time spent speaking and correlation of speaking between users. These observations are used in chapters 6 and 10 as a component of the total system model (considered below). This also represents an interesting approach to analysing activity and interaction based on the affordances of the technology and the user's embodiment as their representative within the computational domain of communication and awareness and may be compared with other approaches such ethnographic observation and conversation analysis (e.g. [Bowers, Pycock and O'Brien, 1996]).

System and Network Behaviour

Chapter 6 builds on the models of user behaviour described above together with detailed observation of the MASSIVE-1 system in controlled situations to construct a predictive model of total network bandwidth requirements for that system. This model demonstrates the potential utility of spatial trading as defined in this thesis to appropriately control and constrain interaction, especially in unicast-based CVEs such as MASSIVE-1.

Chapter 10 develops the simple model of chapter 6 into a more comprehensive and general model of network communication requirements in collaborative virtual environments such as MASSIVE-1 and MASSIVE-2. This model is based on an idealised application model for collaborative virtual environments which is also applicable to general distributed virtual reality systems and to a large extent, to synchronous awareness-based systems based on non-euclidean spaces (if any existed). This model identifies key parameters to characterise user activity and application and protocol behaviour which can be varied to apply the model to different systems and uses. The model defines some key concepts and dimensions which can be used to assess different systems and approaches. The main two are listed below.

- Different systems and approaches to awareness management (or its equivalent) can be characterised and compared in terms of the "accuracy" with which they can represent and realise each user's notional ideal awareness within a given application and environment.
- Systems may also be characterised in terms of the granularity with which replication and communication management is performed. The primary choices are "universe", "world", "region" or "artefact".

Other key outcomes of the analysis of this model are listed below under the headings: state and updates; multicasting; awareness management; and abstractions.

State and Updates

- A distinct phase of state transfer followed by ongoing updates is assumed, at least for participants' embodiments (whatever they may be). It is apparent that the relative importance of state transfers versus updates varies enormously according to the type of application (e.g. the expected patterns of user activity) and the communication media involved (e.g. static graphics, dynamic graphics, audio and/or video). It appears that with richer forms of interaction (i.e. with at least real-time audio) the update bandwidth requirements will dominate in many applications.

- Using repeated heartbeat messages to perform state transfer is only appropriate in extremely limited situations; in such situations an alternative form of multicast state transfer to joining participants would be preferable unless the number of multicast groups available were incredibly limited.

Multicasting

- Using network-supported multicasting can dramatically reduce the total network bandwidth requirements. This is probably appropriate in most situations with the possible exception of high-speed (e.g. ATM-based) LAN-only situations.
- However the use of multicasting may lead to the adoption of less accurate forms of communication management (in order to reduce the number of multicast groups in use and the related management and maintenance overheads).

Awareness Management

- Drawing on the analysis of interest scoping in distributed virtual environments in chapter 2 it is suggested that an appropriate approach to awareness and communication management in CVEs is to use world-based and region-based management at the top level, with individual artefact management being applied to more demanding media such as video.
- It is argued that an explicit model of awareness which includes the effects of context is a good approach to communication management because of its potential accuracy in a range of situations (e.g. both interior and exterior spaces). The spatial model of interaction with third party objects is a model of this type.

Abstractions

- The network requirements model also shows that the introduction of abstractions can reduce not only the load on individual participant's machines, but also the total network load in many situations (because it reduces the total requirements for state transfers). However this depends on its acceptability to users in any given application.

As for the user analysis, above, the form and development of this application model may also be generalised and applied in other situations and to other classes of application. In particular the notion of awareness (which need not be explicit in the application) provides the essential element in forming this model. The same concepts and comparisons might be applied to other domains and styles of system.

This section has presented the key outcomes of this work. The next section is a more subjective reflection on and evaluation of the work performed and presented.

11.3. Reflection

This final section presents more subjective and personal reflections on the work which has been presented and summarised in the previous two sections. Section 11.3.1 describes deliberate or inadvertent omissions from this work (things which might have been done given more time). Equivalently these are possible areas for future work. Section 11.3.2 presents personal reflections on the work which has been done (or otherwise) and the way in which it has been done. Finally, section 11.3.3 tentatively explores something of the "big picture" of the evolving relationship between computer science and society.

11.3.1. Future Work (or What I Didn't Do)

Areas for possible future work are described below in terms of: models of awareness; user and application modelling and testing; network modelling; and system design and implementation.

Modelling Awareness

- It is apparent from use of both MASSIVE-1 and MASSIVE-2 that both the original spatial model of interaction and also the third party object concept itself require explicit representation to the user within the normal interface. As the effects of focus, nimbus and third parties become more unusual (i.e. more "virtual" than "real") then users have difficulty in understanding why they perceive the world as they do or in inferring what they should do in order to achieve particular kinds of interaction. If an approach such as the spatial model of interaction is to be used more generally then it needs to be made sufficiently visible and malleable for "normal" users (whoever *they* may be).
- Similarly, there is a need to make control over interaction more accessible to users (e.g. via manipulation of focus and nimbus).
- The systems and examples presented in this thesis concentrate exclusively on the management of and support for real-time (synchronous) communication and collaboration. These ideas might be applied and generalised to asynchronous situations and to the transitions between synchronous and asynchronous activity (this is one of the emphases of the Aether awareness model [Sandor et al., 1997] which is also inspired by the spatial model of awareness).
- The prototype systems described in this thesis are stand-alone tools which tend to dominate the user's attention. To support more general patterns of use (such as the notion of always being "on-line" but to varying extents) it is necessary to address

direct integration between the awareness model presented here and the other facilities provided by the user's end system. There is also the need to integrate control and use of other individual and group tools within the same context.

User and Application Modelling and Testing

- While the motivational focus of this work has been on the support of potentially large numbers of simultaneous users it has not been possible to perform large-scale trials within the temporal and resource constraints of this work. This is a clear area for ongoing work and is necessary to address the potential utility of the system designs and the network traffic model. Also, and perhaps more importantly, large-scale tests will be needed to assess the effectiveness of the spatial model, third party objects and the CVE approach and philosophy in general.
- With regard to capturing and characterising user activity and behaviour there is a need to consider: larger user populations including, for example, "expert" and "novice" users; a range of application types beyond small group tele-conferencing as considered here; a range of interface metaphors and technologies, including the use of immersive displays and tracking technologies. All of these are likely to have a profound effect on the type and pattern of user activity, which in turn impacts system and network requirements.

Network Modelling

There are a number of well defined extensions to the network requirements modelling of chapter 10 which could form areas for future work. These include the following:

- Consideration of non-trivial network topologies. The current model considers *total* network bandwidth only, i.e. treating the network as a single shared broadcast segment. The model might be extended to consider, for example, distinct LANs within a broader WAN, each with different bandwidth and/or cost characteristics.
- The model might be developed to explicitly address client-server style implementations (the current model focuses on peer-to-peer communication). This may also include consideration of user access over low-bandwidth tail-links (e.g. modems). This would go beyond the analysis of Funkhouser [1996] in considering client-server traffic as well as the inter-server traffic considered in that analysis.
- The model is currently for long-term average bandwidth only. There is a clear need to model and reason about transient effects such as possible peaks in bandwidth requirements associated with state transfer operations. This is particularly important in the context of low-bandwidth connections such as narrow WAN or modem connections and for provisioning of networks and services.

System Design

- As noted above it has not been possible to test these systems with as many participants as might have been hoped. From preliminary performance tests it appears that the performance of individual participant's machine will be a key limiting factor even with relatively small number of mutually aware users. There is significant future work to be done to characterise the actual limitations and to explore ways of increasing total system scalability, for example by adaptive control of interaction, or by off-loading activities (such as elements of audio mixing and management) to external shared services.
- Flow control and congestion management is largely neglected by the current systems. This is an open research issue for multicast communication in general. It is complicated in this application by the use of several multicast groups in parallel, with continually shifting interests. It would also be interesting to consider the ways in which back-pressure might be represented to or applied to the individual participants in a CVE so that they could reason about the observed behaviour of the system and modify their conduct accordingly.
- One possible application of CVE technologies in the long term is in the consumer market as a form of "Inhabited TV" in which viewers can "step inside" their TV set and join the audience or even the main content of the show. This is also considered in section 10.3.3. This implies a need to move gracefully between very different modes of participation such as passive observer, mutually aware audience member, or central participant. The metaphors, interface technologies and system support for these varying modes of participation and the movement between them remain areas for future work.
- In a similar way the movement between synchronous or foreground use of the system and asynchronous or background use of it (as a tool for awareness) involves similar transitions between modes of participation.

System Implementation

This section lists a number of obvious extensions to the implemented systems, in particular MASSIVE-2 which is still being used and developed (unlike MASSIVE-1).

- Communication management in MASSIVE-2 is based on disjoint worlds and on spatial third party objects (i.e. regions). An interesting area of future work would be to extend this management framework to handle logically defined groups (e.g. artefact hierarchies or security-related groups) and also to integrate per-artefact communication management in appropriate cases. This thesis has argued that high-cost media such as video and perhaps also audio should be managed on a per-artefact basis because the cost to a participant of inaccuracy is so high. However the system does not currently support this.

- The current systems supports text, audio and 3D graphical communication but not real-time video. This is an area of ongoing work being explored by Gail Reynard in the Communications Research Group as part of her PhD studies.
- In normal working situations it will be appropriate to integrate the CVE system with the other tools which the user makes regular use of such as World Wide Web browsers, word processors and email facilities. This is not currently supported by either system.

11.3.2. Personal Reflections

This section gives brief personal reflections on the work presented in this thesis and the activities which lie behind it. The aspects considered in turn are: "how it went", "how it turned out" and "what should have been done differently".

How It Went

I (the author) think that the work presented here started particularly well, with a first version of MASSIVE-1 being usable about six months into the period of study. This was important because the overhead and time requirements for obtaining useful experience with the system were unexpectedly demanding. The timely occurrence of the BT/JISC funded Inhabiting The Web project allowed the system to be used to hold 17 wide-area meetings and provided extremely useful experience backed up by activity and traffic logs.

A large amount of time was consumed by aborted developments of intermediate versions of the successor to MASSIVE-1. I was unable to locate a distribution platform with which I felt happy with and which was appropriate to the structure of the second system. This was also complicated by the relatively late completion of the network analysis of section 6.3.6 which demonstrated the importance of network supported multicasting for reducing total network bandwidth requirements. With a target of 100 simultaneous users it became clear that multicasting would have to be employed. This further reduced the choice of suitable distribution systems and I decided eventually to create a new distributed object system which integrated multicasting in a natural and relatively light-weight fashion.

In parallel with the development of the core system facilities I was also reflecting on how the constraints of multicasting (i.e. common delivery to a set of observers) might be reflected in an awareness model and vice versa. Consideration of objects as representative of multicast groups was the basis from which the third party object concept was refined and developed during the first half of the third year of study. The completion of MASSIVE-2 was pushed back towards the end of the third year of study, and this was exacerbated by beginning to lecture three months into the third year.

There has not yet been time to exercise MASSIVE-2 in the way that MASSIVE-1 was in the ITW trials. However the NOWninety6 poetry performance using MASSIVE-2 (see section 9.2.1) was a significant motivation (understatement) to get the system operational, reliable and user-friendly. My effort has been diluted over the final year of study by teaching and other research commitments and I have concentrated on incremental improvements to the system, together with the evolution of the analysis and evaluation presented in chapters 9 and 10 in particular.

My working style tends to be one of rapid prototyping and exploration in parallel with conceptual work, modelling and analysis of existing systems. This is apparent in the evolution of the third party object concept alongside the development of MASSIVE-2. I find that this can give a richer (and more interesting) development process with a more realistic "dialogue" between theory and experience.

How It Turned Out

MASSIVE-1 was explicitly a minimal prototype implementation of the spatial model of interaction. As such I think that it has been very successful. It has provided a demonstration platform to explore the concepts of the original model. It has also been useful for holding distributed meetings over wide area (though high-bandwidth) networks. It has been particularly satisfying to see others deriving new insight from its use and enjoying involvement in virtual meetings.

MASSIVE-2 was a more ambitious system intended to be a general CVE and application development platform. I consider it to have been partially successful. For example, it successfully hosted the NOWninety6 poetry performance. As an example of a different class of application the WWW-3D collaborative web-browser (see section 9.2.4) has been ported to MASSIVE-2 from the DIVE system [Hagsand, 1996] and region-based enhancements have been made to it. Other members of the Communications Research Group at the University of Nottingham are also developing their own work with and within the system. However, preliminary performance tests suggest that the system will support approximately twenty mutually aware users on typical current workstations (e.g. Silicon Graphics O2); this is still somewhat short of the design goal of 100! Likely reasons for this appear to be: the relatively heavyweight character of the distributed object system, especially when handling continuous media (e.g. audio); and the current focus on peer-to-peer communication, with little facility to off-load or share computation with other machines in the network. The support for abstractions should address this second issue to some extent, but I think that it will need more dynamic and flexible management than exists at present.

I am very happy with the third party object concept. In particular, it has integrated with the original spatial model of interaction in a very elegant way. It has also demonstrated at least some of its flexibility in MASSIVE-2. The notion of using (negotiated) awareness to control the third party object appears particularly promising and expressive. MASSIVE-2 is currently rather limited in this respect, concentrating as it does on spatial membership controlled third party objects (i.e. regions), however there appears to be significant unrealised potential within the model that might be used in future systems.

What Should Have Been Done Differently

I regret the time which was effectively wasted on aborted versions of MASSIVE in the middle of this work. I would ascribe this to a lack of early modelling and performance evaluation, and also to a general tendency to start prototyping a little too soon.

In principle and in retrospect I would like to have found an existing distributed system platform to use, or possibly used an existing virtual reality system or toolkit such as DIVE or MRToolkit [Shaw et al., 1992]. This should have allowed me to concentrate more on the individual issues and areas which I was exploring. However, I doubt that this would ever have happened. The areas being explored deal with relatively deep aspects of distribution and communication management which require very low-level support. Also, I think (or rationalise?) that an intimate familiarity with all levels of the system gives a more intuitive sense of the "aesthetics" of distribution. I hope that this will have created a more harmonious and integrated system rather than an awkward simulation of the desired effects running over a reluctant infrastructure. Finally, I may never have the time to indulge myself quite like this again.

The last thing which I would have done differently is more of a strategic error: in both systems I attempted to use a common distribution system for all interaction and communication, from RPCs through graphical updates to audio streams. This has certain advantages in constructional and conceptual simplicity. However it is also one of the main reasons why I believe the scalability of MASSIVE-2 is currently limited. It is not so much that the *metaphors* are wrong, but that streamed media in particular are both highly demanding and also amenable to medium-specific handling and optimisations. For example, the encoding rules for audio and video are entirely medium specific. To have an additional generic layer of marshalling (as there is at present) is a real waste. Also, the despatch mechanism appropriate for object oriented programming (as used in MASSIVE-2) is much more heavyweight than is required for streamed media in most situations. In MASSIVE-1 part way through development the audio medium was split out into a separate generic audio service with tailored communication protocols. I think that the same thing should be done with MASSIVE-2. Similarly, if video is integrated it will need careful and appropriate handling (for reasons of performance, rather than correctness).

11.3.3. The Big Picture

This thesis concludes by considering the way in which the areas covered in this thesis may be viewed within the broader picture of computer science and society. This is considered it terms of: inter-disciplinary research, inhabited television and ubiqitous computing.

Inter-Disciplinary Research

Inter-disciplinary research appears to be extremely difficult to maintain, with barriers of language and terminology, of philosophy and of ideology, adding to the more mundane logistical and administrative barriers which exist between departments and organisations. One of the interesting aspects of the work presented in this thesis is the way that it brings together concerns from sociology and psychology together with issues of computer science, from CSCW and CHI to distributed systems and networking. The motivations behind CVEs derive in part from ethnographic observations of everyday work practice. Within the lifetime of the work presented here this has come full circle, with these same ethnographic techniques being applied to environments which they have inspired (see in particular [Bowers, Pycock and O'Brien, 1996]). These observations yield new insight and understanding and in this way practice progresses. When inter-disciplinary research can be made to work it can be exceptionally productive and innovative. Perhaps one way in which it can be made to work, as in CVEs, is around a common ideal or driving application. Furthermore, within the framework of "Inhabited Television" (see below) artists, writers, producers and directors all become potential collaborators. It is to be hoped that this spirit (and reality) of collaboration will be maintained and enhanced, both within this area of work and also more generally. Computer science is not an end in itself: the highest duty of computer science is in the service of man (or the glory of God, depending on your perspective).

Inhabited TV

If one were to adopt a visionary perspective then perhaps the "killer application" for collaborative virtual environments would be "Inhabited Television", i.e. making this type of technology and environment available to domestic users in their own homes. The traditional model of television is one of highly centralised production and coordinated large scale distribution to large numbers of passive viewers. This approach is required by the limitations of the technologies traditionally used for television, i.e. high-investment terrestrial broadcasting. The emerging area of *Interactive* TV (see for example [Salmony, 1995]) supplements this unidirectional flow of content with a (typically very low bandwidth) reverse channel, allowing simple feedback from individual viewers to the content provider. Uses for this back-channel typically include requesting content (e.g. video-on-demand) and simple responses to content such as voting and tele-shopping (ordering goods and services). *Inhabited* TV seeks to extend this model in two significant respects: firstly it introduces direct communication between the distributed viewers; and secondly it expands the size and nature of the back-channel to support richer interaction and moment-by-moment involvement with the content. That is, it allows isolated *viewers* to become involved *participants*: participating in a collective (mutually aware) audience and being able to participate directly in the content itself, for example by making a significant contribution to it.

Moving from CVEs as they exist today to the notional future of Inhabited TV presents many major technical and social challenges. For example, the issue of scalability, which has been one of the concerns of this thesis, assumes massive (!) proportions. In the UK alone peak viewing figures for conventional broadcast TV regularly exceed 10 million simultaneous viewers. The global viewing figures for the funeral of Diana, Princess of Wales, in September 1997 are reported to be in excess of 2 billion. There is a long way to go to address this scale of use. There are also other issues such as providing users with effective navigation and rich interaction within a relatively unconstrained (and often social) domestic environment. Also, viewers will need to be able to move between various modes of (non-) participation, for example from passive viewing, through co-aware audience membership, to central participation. There are also profound issues relating to the content of Inhabited TV. What should it be like? How should it be structured? Who will create it and how? Who will control and manage it? And so on.

Mixed Realities and Ubiquitous Computing

Bowers, O'Brien and Pycock [1996] make the point, in observing the process of staging a meeting in MASSIVE-1, that only part of the activity and communication is occurring within the system, i.e. within the *virtual* world. Each participant is still very much part of their own physical environment such as their office or their home. There seems to have been a tacit assumption in much research related to virtual reality that *ideally* people will "leave the physical world behind" and step whole-heartedly into the virtual, at least for the duration of use. This is clearly not usually the case. Nor is it even ideal except in relatively constrained training and entertainment applications. It is necessary, rather, to consider the real and virtual worlds evolving and coexisting in parallel at all times.

This may also be seen as a perspective or approach to the philosophy of ubiquitous computing [Weiser, 1993], i.e. the notion that everything, including normally mundane and everyday objects, should not only be computerised but also communicating and cooperating in a benign web of natural electronic assistance. Ubiquitous computing can be seen as breaking down the boundary between "real" and "virtual" (i.e. electronic) by infusing the real world with the virtual by technological means. Similarly, the techniques of augmented reality (as in [Bajura et al., 1992], for example) seek to supplement or overlay the real world with virtual artefacts and information. One might say that the ideal is not *virtual* reality, but *mixed* reality, a merging of real and virtual objects and of real and virtual spaces. This is an area which has begun to be explored by the author and others (see for example [Milgram and Kishino, 1994] and [Benford et al., 1996]). These approaches and ideas may have a profound influence both on the way in which these technologies evolve and ultimately upon the way in which technology and perhaps even reality are understood by society at large.

Appendix A.
User Profiling

This appendix presents a quantitative analysis of aspects of user behaviour in MASSIVE-1. This allows an approximate model of expected user behaviour to be built. Such a model is a key element in assessing the network and computational resources required to support varying numbers of users in any CVE system and is used in the network traffic models of chapters 6 and 10. Section A.1 begins by describing the sources of data which have been utilised. Section A.2 then goes on to present results for user movement while section A.3 considers the use of audio. The data presented here is based on the last six meetings held within the ITW project. These meetings were relatively free of technical problems, and more data is available for these than for other meetings. Consequently the results presented are derived from a relatively limited class of CVE usage, i.e. for small structured meetings which are dominated by inter-personal communication. None the less, it can provide a starting point for analysis and modelling in other application areas.

Further details of these trials can be found in [Greenhalgh et al., 1997].

A.1. Data Sources

There are five main sources of data available about the use of MASSIVE-1. These are:

- log files generated by the MASSIVE-1 user client programs;
- network traffic logs captured using the UNIX utility *tcpdump*;
- videos of the meetings from the perspective of one or more participants;
- questionnaires completed after many of the meetings held within the ITW project; and
- personal reflections on the meetings from those involved.

The user-oriented evaluations of chapters 5 and 9 are based primarily on personal reflections and discussions with those involved, supplemented by some of the video material captured during meetings. In this appendix use has been made of the more quantitative data from the first two sources - log files and network traffic. The questionnaires were set and analysed by others involved in the ITW project and are not dealt with directly in this thesis.

The next two sections provide additional information about the scope of the log files and network data used, before moving on to consider aspects of movement in virtual worlds.

A.1.1. MASSIVE-1 Log Files

Quite early in MASSIVE-1's development facilities were added to all programs, though especially the user client programs (the graphical, audio and text clients) to generate time-stamped logs of key events. The events which can be recorded include:

- starting the application;
- moving to a new virtual world (through a portal);
- moving and changing orientation within the virtual world;
- updating the graphical view (i.e. rendering a single frame);
- changing focus and nimbus settings;
- making graphical gestures;
- speaking (or more specifically, whether the user's graphical "mouth" is displayed); and
- sending and receiving network data.

Each program generates its own independent log file, and one of the problems of analysing this information is relating the information from different programs, especially since each machine's system clock time is typically different.

Explicit user actions are required to make MASSIVE-1 applications generate event logs. The event logs are also generated on each user's local machine. Consequently, it has not been possible to get complete sets of log files for all meetings.

A.1.2. Network Traffic Data

In addition to the high-level and network traffic information recorded directly by MASSIVE-1 applications, the *tcpdump* program was also used to capture records of local network traffic for the audio service and for the server components such as the collision manager, trader and world servers. These records include packet type (e.g. UDP, TCP), size and IP source and destination addresses for packets which are observed on the network (shared Ethernet). However, being a general purpose utility, it cannot provide the kind of application-specific information that can be obtained from MASSIVE-1's own event log files.

Network traffic logs are available for many of the meetings held within the ITW project for portions of packet audio data and for traffic to and from the server components.

A.2. Movement

The area of user movement has proven to be the most amenable to analysis based on the available data. Specifically the MASSIVE-1 event logs from a user client application record that user's moment-by-moment position and orientation within the virtual world and also any change of world (due to passing through a portal). This analysis begins with an overview of the movement data available. This is then used to address three issues for the meetings under consideration. The things which will be established are:

- the fraction of time which people spend moving rather than stationary, and whether participants move simultaneously or independently;
- whether participants move through portals singly or in groups, and when moving as groups how long the combined transition takes;
- whether participants return to worlds which they have already visited, and if so after what period of time.

These have implications for, respectively:

- network and computational requirements for handling movement within virtual worlds;
- the scope for using multicast to handle inter-world transitions by groups of participants;
- whether world caching would be effective and if so on what time-scale.

Figure A.1 on page 202 is a visualisation of movement-related information from the ITW meeting held on the 25th September 1996. Time is shown along the horizontal axis, while a combination of participant and world is presented on the vertical axis. The plot is divided into six horizontal bands which correspond to the six virtual worlds which were visited by participants during this meeting. The top-most band corresponds to the "presentation-world", the next to "balloon" world, and so on. Each participant is allocated a consistent slice of each of these bands; when a participant is present in a particular world this is indicated by a horizontal line in that participant's slice of the world band. A thick line or box indicates that the user is moving whereas a thin line indicates that they are stationary. When participants change worlds this is represented by vertical dashed lines between the worlds in question. For example, consider the top-most band of figure A.1 on page 202 which represents "presentation-world": participant number 3 jumps from "meeting-world" to "presentation-world" at approximately 15:10, moves about for a short time (probably checking if anyone else is there) and soon returns to "meeting-world".

Fig.A.1: visualisation of movement data from the 25th September 1996 ITW meeting

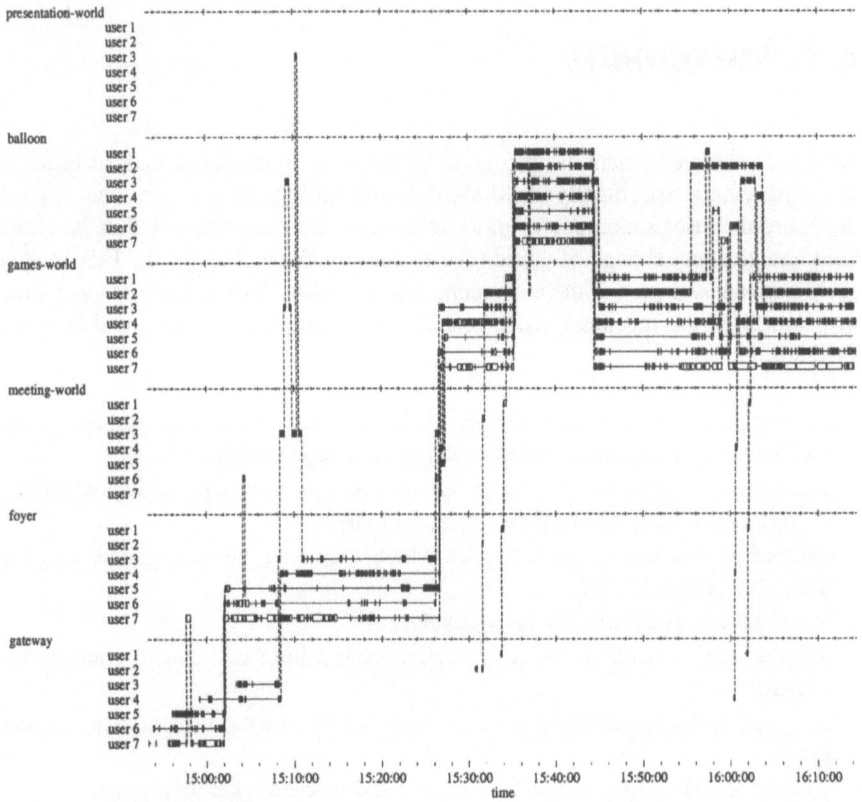

A.2.1. Time Spent Moving

Figure A.2 on page 203 shows a graph of the percentage of time present in a world which participants spent moving rather than stationary. Each participant was considered independently, and for each visit to a world the time for which they were moving and the total time for which they were present was established. In figure A.2 on page 203 the horizontal axis corresponds to the percentage of time spent moving, while the vertical axis indicates the number of participants-seconds for which that level of movement was observed. Both cumulative and point distributions are shown with 1% bucket sizes.

The average percentage of time spent moving for all participants and worlds is 19.6%, though it is clear from figure A.2 on page 203 that this measure is highly variable. Independent of world visited the figure for each participant varies from 7.2% to 28%. On the other hand, for each world averaging over all of its visitors the percentage of time spent moving varies from 7.5% (in the classical music world of the ITW end-of-project virtual party) to 54.6% (in the disco world, home of the disco dancing competition); in the main meeting world the average value was 16.7%.

Fig.A.2: percentage of time present in a world which was spent in motion

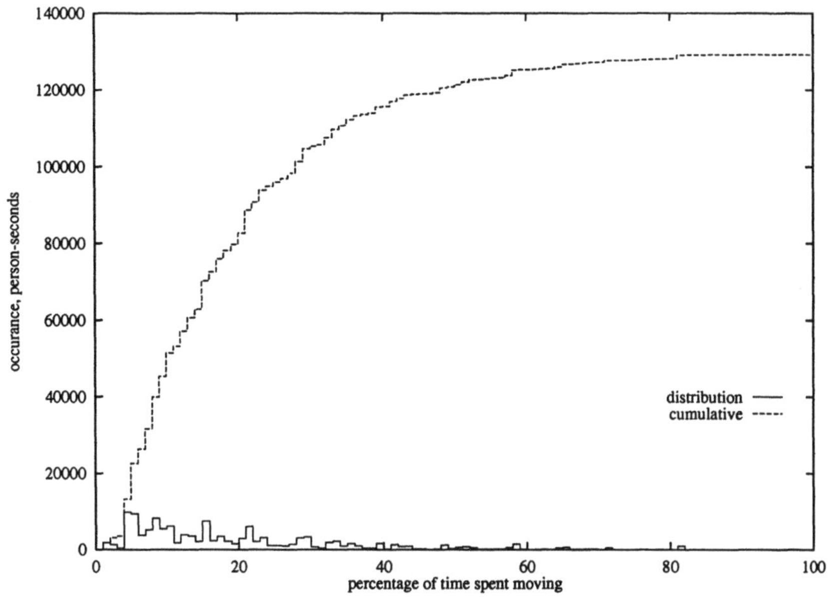

The figures above are based on analysing each participant independently. In addition each world was analysed over the duration of the meetings to determine whether participants tended to move at the same time, or independently. Figure A.3 on page 204 shows the distribution of the number of participants in the same world at the same time who were moving. The solid line shows the observed distribution, while the dashed line shows the distribution that would be expected if all participants ignored each other and just moved when they felt like it (based on the same overall time spent moving, as previously determined). The dotted line shows the total world population distribution, i.e. if every user moved all of the time then this curve would result. It is apparent from the graph that participants do coordinate their movements to some extent - the deviation from random activity is significant at the 99.9% level. However the overall shape of the graph is very similar, and large numbers of participants moving simultaneously is possible - but less likely - in either case.

As noted in the introduction to section A.2, the time which participants spend moving and the amount of correlation between participants will have implications in terms of the network and computational requirements of communicating, processing and presenting each participant to the rest of the virtual world. Specifically, when a participant is moving, position and orientation updates need to be sent over the network and received and processed by each observing process. When dealing with

Fig.A.3: distribution of numbers of virtually collocated participants moving simultaneously

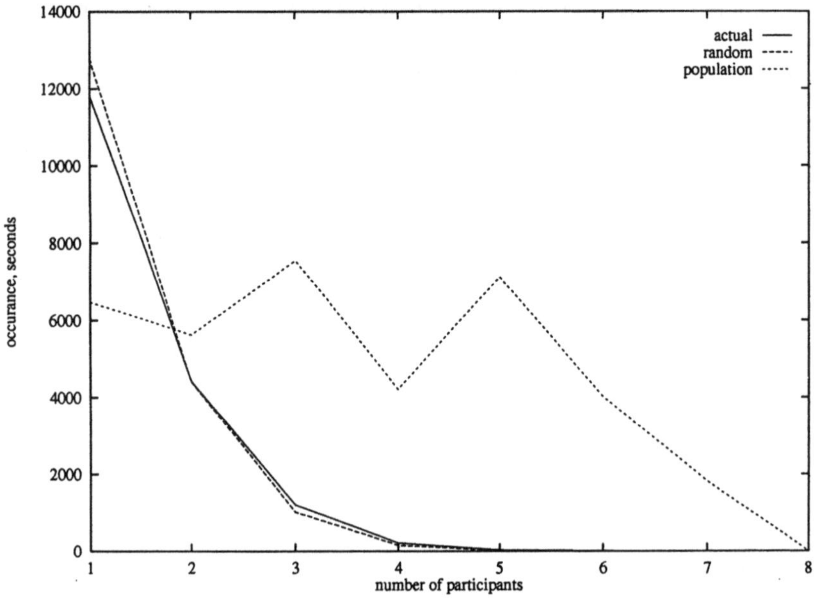

non-deterministic agents (e.g. people) techniques such as dead reckoning may - or may not, depending on the application - reduce the number of such updates, however they can never entirely eliminate the need for some of them. But once a participant is stationary no further positional updates are necessary until they start to move again. Consequently the amount of time which participants spend moving can be a significant factor in assessing network and computational requirements.

The overall average percentage of time spent moving observed in the six meetings being analysed here was 19.6%, as noted above. This may be taken as a base-line value. However there is also a large variation between individuals and also a significant task dependence, as indicated by the differing averages in different worlds. Consequently, the figure arrived at here should be treated as something of a rule of thumb, rather then a definitive answer. Additionally, when considering the combined instantaneous load due to several participants this analysis shows that it is not valid to assume that their activities will be independent and uncorrelated. Rather, at least when participants are involved in a common task, there is a small, but statistically highly significant, element of correlation between their activities. This argues for additional caution when considering, for example, the scope for exploiting statistical multiplexing of movement-related traffic for larger numbers of users (i.e. being able to require or reserve less bandwidth on the basis that while some users are moving - and generating network updates - many other users will not be).

A.2.2. Group World Transitions

Having considered one key aspect of movement within a world this analysis now considers two aspects of moving *between* worlds. First it considers the likelihood and form of coordinated inter-world transitions by groups of participants and then whether and in what circumstances participants return to previously visited worlds.

It might be expected from some of the activities organised in the meetings being analysed that group world transitions would occur. For example, participants would typically gather initially in the gateway world and wait for others to arrive. Then at some point the meeting organiser would invite everyone to go through to the meeting world for the formal start of the meeting and all of the participants would move - in a vaguely coordinated fashion - through to the meeting world. The purpose of this aspect of the analysis is therefore not simply to discover whether such transitions occur. Rather, the purpose is to assess the significance and character of such transitions. Once the data has been presented its significance will be considered.

For the purpose of the automated analysis a group world transition is defined as an event in which two or more participants who are in a world at the same time move via a single portal jump to another world so that they are together again. Figure A.4 on page 205 shows the incidence of singleton and group world transitions in the meetings analysed. The solid line shows the number of incidents, while the dashed line shows the total number of participants involved in those incidents, e.g. each group transition *incident* for group size four must have involved four *participant* transitions.

Fig.A.4: distribution of group size for group world transitions

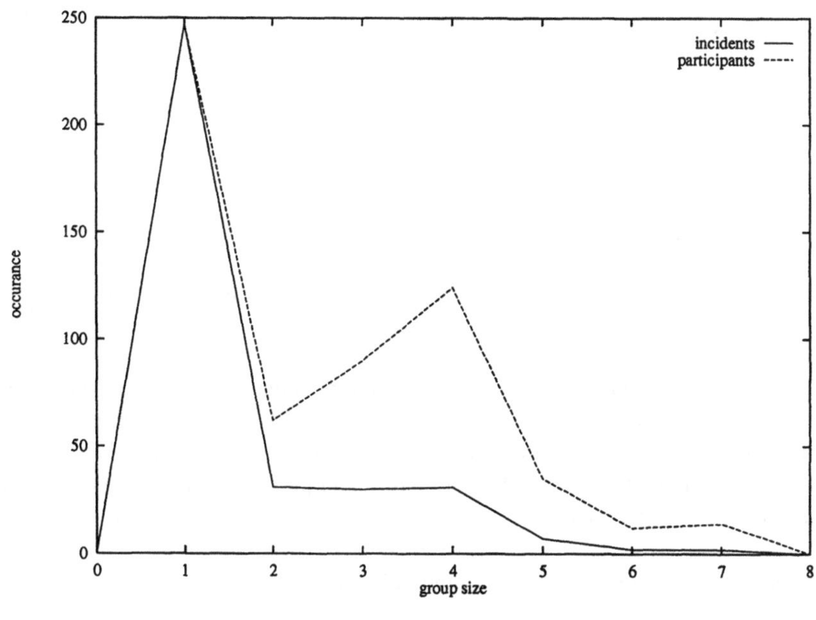

Summarising the underlying data:

- participants jumped to new worlds on a total of 584 occasions;
- of these, 337 (58%) were in groups of two or more;
- individuals or groups made world transitions on 350 occasions;
- of these 103 (29%) were group transitions;
- the average size of those groups was 3.27 participants.

Figure A.5 on page 206 shows the distribution of world entry delay for participants involved in group transitions. For each member of a group (excepting the leader) it shows how much time elapsed between the group leader and the group member reaching the destination world. The range of delay shown on the graph, up to 30 seconds, accounts for 203 (87%) of the 234 non-leading participants to make group world transitions. 104 (44%) of these occur within 5 seconds, while 159 (67%) occur within 10 seconds.

Fig.A.5: distribution of group member arrival delay for group world transitions

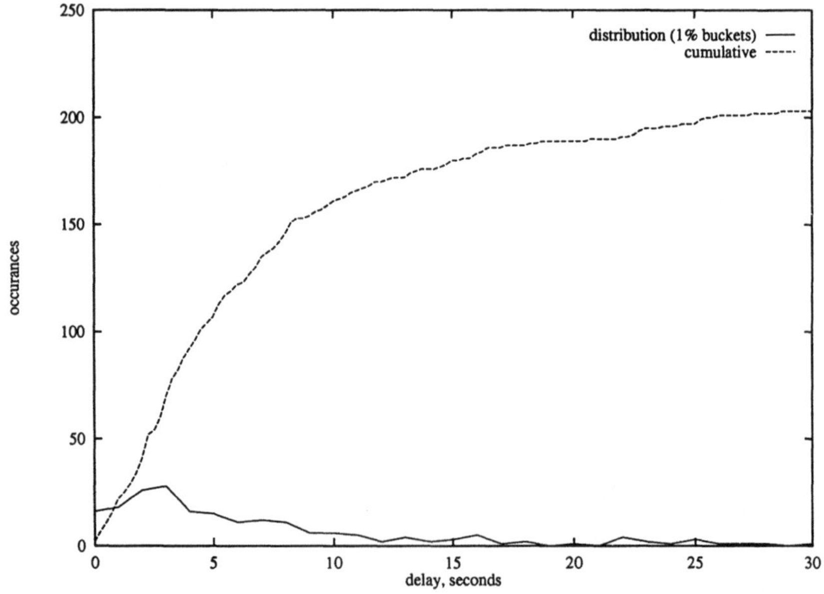

Group world transitions are important when considering the design of and requirements for CVEs because moving to a new world is a significant event which will almost always involve an exchange of data between the participant's process(es) and the rest of the system. In MASSIVE-1 for example, on entering a new world the user's client applications will:

- terminate connections to objects in the old world;

- be informed of the existence and identity of nearby objects in the new world;
- establish network associations with each; and
- exchange general and medium-specific information such as location, awareness, name, graphical appearance, etc.

This can result in a significant but transient burst of network traffic and a corresponding load on other processes. Different systems will organise this information in different ways and obtain it from different sources, but there will still be some requirement for the participant's application(s) to learn about the new world. This makes the occurrence of group world transitions important in two respects:

- a coordinated movement by a large group could generate a much greater transient load than might be expected if inter-world movement were assumed to be independent and uncorrelated; and
- specialised system support for group world transitions (for example, based on the use of network-supported multicasting of new world information to all group members) could both alleviate this problem and also reduce the total network load relating to world transitions when compared to a model of independent and uncorrelated movement.

For example, for the meetings analysed, imposing a world transition delay of 10 seconds would both ensure that state transfers for the same world did not need to be performed more than once in any 10 second, and would require approximately 325 unicast and 100 multicast state transfers rather than 584 unicast state transfers. Imposing longer delays on world transitions would increase the effectiveness of group transfers by allowing more transitions to be grouped whereas shorter delays would include fewer transitions.

Before moving on to consider participants returning to worlds a little must be said about the general applicability (or otherwise) of this result. As was noted at the beginning of this section group world transitions were an organised aspect of the activities being analysed; will they occur in other applications and situations? Such an question cannot be answered definitively without gathering a great deal more data about a wide range of different applications and scenarios. However some more subjective and tentative observations can be made:

- MASSIVE-1 is not alone in adopting a multiple world model with portals between worlds (see for example DIVE [Carlsson and Hagsand, 1993] or in a more limited sense dVS [Grimsdale, 1991]);
- the same effects and results would apply for systems structured using regions (as in MASSIVE-2 and Spline [Barrus et al., 1996]) as for worlds (but at the granularity of regions);
- the world designer for the ITW meetings (Adrian Bullock) chose independently to structure the meeting space as a number of different worlds with tailored content and form; and

- the world and portal model was widely accepted and effectively employed by participants.

It may be argued from these observations that a multi-world structure (or an equivalent regionalised structure) is a generally useful and appropriate virtual design style. So it may be anticipated that group world transitions will occur in many applications involving formal or informal cooperation and interaction, for example, as common interest groups form and dissolve or as time-linked activities such as performances and meetings begin and end.

A.2.3. Returning to Worlds

The final aspect of participant movement which will be considered in this analysis concerns the incidence and character of return visits to virtual worlds, i.e. when a participant visits the same virtual world on more than one occasion during the same meeting.

Figure A.6 on page 208 shows the cumulative distribution of time elapsed between consecutive visits by individual participants to any world. There are a total of 353 return visits in the data analysed compared with 231 first visits. The average time lapse between leaving and re-entering the same world is just over 7 minutes, while half of these return visits occur within 3 minutes.

Fig.A.6: distribution of time elapsed between consecutive visits to the same world

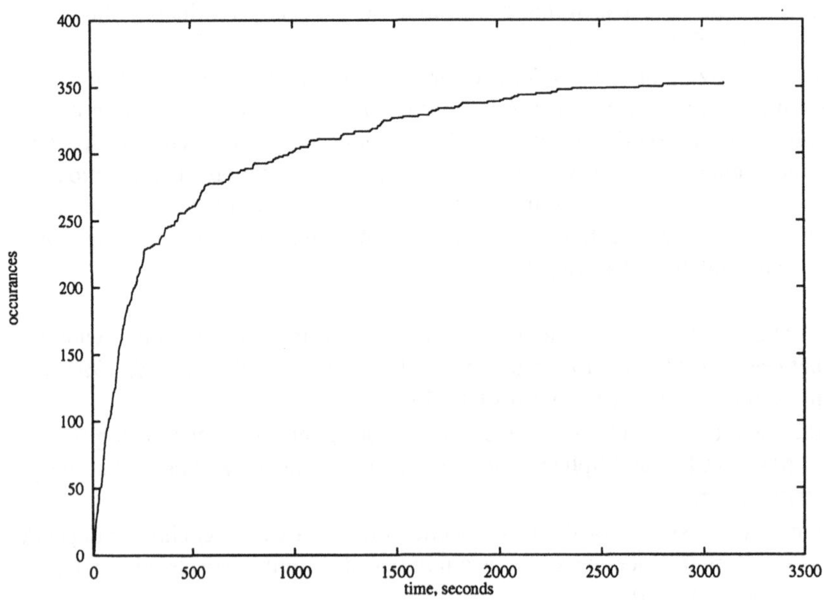

The occurrence of revisiting worlds is important for the same reasons as group world transitions: because changing worlds is a significant event which implies an exchange of information with corresponding requirements for network and computational resources. In particular, world return visits are important because of the possibility of world state caching in (or near) a participant's application(s). For example, for the data analysed, if each user client maintained a cache of world state for at least 6 minutes after visiting a world and if the worlds did not change on this time-scale then approximately 40% of world state transfers could be satisfied from the cache.

The above example of the potential utility of world state caching must be qualified in two areas: the extent to which return rates can be generalised from the specific meetings analysed and the extent to which worlds will remain static over the times periods in question (and how this can be established by the applications).

Dealing with these issues in turn, it is clear that if return visits were more common or occurred sooner then caching would become increasingly effective and conversely, if return visits were rarer or more delayed then caching would become less effective. As with group world transitions it may be argued that most world return visits in the meetings considered were peculiar to the style and organisation of these meetings and cannot be generalised to other applications. However there is a counter-argument that these worlds and meetings were not established to demonstrate these effects, but that these effects emerged from natural choices of world and meeting structure which may recur in other systems and applications. For example, the worlds for the ITW meetings were structured (for the most part) in a hierarchy which is a common method for organising related objects. A natural consequence of this is that the worlds closer to the root of the hierarchy were visited more frequently and were returned to more often, for example by participants in the process of moving from one task-oriented world to another. So revisiting of worlds may be a general effect.

In MASSIVE-1 worlds the background content is static while the participants are highly dynamic. In other systems and applications the differences may be less clear cut. In any case, for caching to be worthwhile there must be significant elements of world state which do not change between visits and there must be some well defined method of establishing what has not changed and of efficiently combining cached and new information. This is left as an exercise for the reader.

A.3. Audio

In addition to the analysis of movement and world transitions presented above there has also been an analysis of audio-related activity in the same six MASSIVE-1 meetings. The information that was available relating to participants' audio activity was:

- records of audio data packets captured in the network traffic log; and
- records of when the visual "mouth" was visible from the MASSIVE-1 user client log files.

Each of these sources of data corresponds to a simple threshold test of audio volume. The visual mouth appears when the sound captured by the participant's microphone exceeds an experimentally chosen level, while the audio service sends audio packets (which are recorded in the network traffic log) when the sound captured by the microphone exceeds a lower experimentally chosen level. In general, the audio server errs towards treating silence (or rather background noise) as speaking, while the "mouth" errs towards treating speaking as silence; this is intended to encourage participants to speak up based on (not) seeing their own "mouth".

These sources of data are considered in turn before drawing joint conclusions.

A.3.1. Network Audio Data

The data for logged network audio traffic includes all audio in the same world as the principle meeting organiser. It extends to 33198 person-seconds of audio from 125571 person-seconds of apparent presence, i.e. the average speaking proportion is 26.44%. Apparent presence of audio participants is deduced from periodic audio timing packets which are sent by the audio service even when full audio data is not being transmitted. Figure A.7 on page 211 shows the number of simultaneous speakers for the data set (solid line) and the expectation if speaking was an independent event with no correlation between virtually collocated participants (dashed line). The dotted line shows world population; if all the participants spoke continuously then this curve would result. The observed distribution differs from the uncorrelated case at the 99% significance level, but the difference at each data point is small - the difference is only significant because of the very large sample size.

A.3.2. Visual "Mouth" Data

The data concerning the visual presentation of a "mouth" representation (actually more of an abstract speech balloon) is available where participants have correctly enabled MASSIVE-1's own logging facilities and have returned the event log files. The total participant time covered, 129550 person seconds, is similar to the network audio data. However the amount of time for which the mouth is shown is much less: 9355 person seconds, or 7.22%. Figure A.8 on page 211 shows the number of simultaneous "speakers" (actual and uncorrelated expectation) and world population for this data set. As for the network audio data the deviation of the observed number of simultaneous speakers from an uncorrelated distribution is significant at the 99% level, but small in percentage terms. Complete event logs for all participants are not generally available, and this is reflected in the shift of the world population distribution towards the left (fewer simultaneous participants) when compared to the network audio data. The distributions for speaking drop off much more steeply than for the network audio data because of this but primarily because of the much smaller amount of speaking recorded.

Fig.A.7: correlation of network audio packets

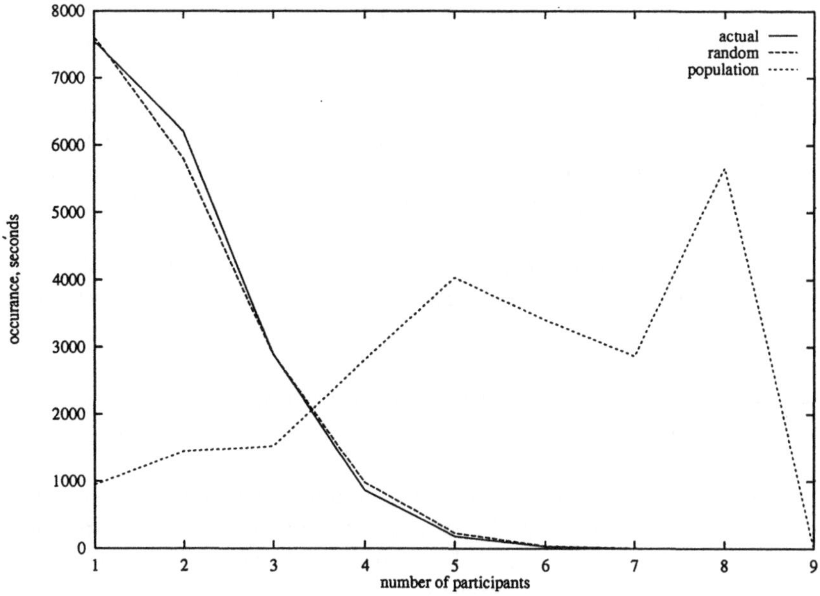

Fig.A.8: correlation of visual "mouth" data

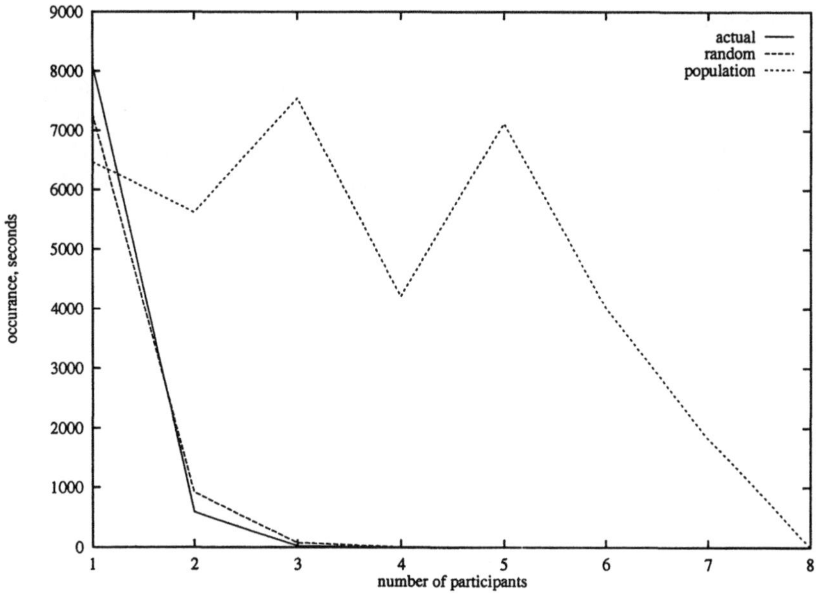

A.3.3. Audio Data Conclusions

Before the analysis it was anticipated that the data for speaking might exhibit significant anti-correlation, reflecting the assumption that one participant would be less likely to speak when another participant was already speaking. In fact this was seen to a limited extent in the visual "mouth" data, but not at all in the network audio data. At first sight this might suggest that the participants are ignoring one another, or that the network delays might cause a break-down in normal conversational turn-taking. However a subsequent comparison between segments of the logged data and the video recordings revealed a more mundane reason: much of the apparent "speaking" was actually due to background noise (e.g. typing, other activities in the office, breathing noise) and feedback from open speakers. A number of observations and reflections can be made:

- Use of open speakers (at least without sophisticated echo cancellation techniques) can increase the apparent number of speakers and the resulting network load by replaying sounds back into the system.
- Using more aggressive silence detection algorithms might reduce network traffic (moving the 26% speaking rate for network data towards the 7% speaking rate for the "mouth"), but at the risk of missing significant quiet utterances and non-speech noises (e.g. hmms, grunts and ahs).
- Whatever the participants "real" behaviour in terms of speaking (as determined by a human expert for example) it is the behaviour that may be deduced automatically (as here) that is significant for assessing network and computational requirements.

This concludes the analysis of quantitative aspects of user behaviour in the MASSIVE-1 meetings. This forms a key component of any model of network and system requirements for CVEs, and elements of this analysis are used in chapters 6 and 10 which develop network traffic models for MASSIVE-1 and MASSIVE-2, respectively.

Appendix B.
Exploitation and Dissemination

This appendix lists publications and projects to which the work presented in this thesis has already contributed. Section C.1 lists publications which are significantly or predominantly derived from this work. Section C.2 lists publications which include more limited aspects of this work. Section C.3 identifies projects which this work has contributed to.

B.1. Primary Publications

The analysis of scoping of interaction in distributed virtual reality systems from chapter 2 has been published in essentially this form as:

> Greenhalgh, Chris, and Benford, Steve (1997), "Boundaries, Awareness and Interaction in Collaborative Virtual Environments", Published in the *proceedings of the 6th International Workshop on Enabling Technologies: Infrastructure for Collaborative Enterprises (WET-ICE)*, June 18-20, 1997, Cambridge, Massachusetts, USA.

Preliminary development of MASSIVE-1 was published as:

> Greenhalgh, C. (1994), "An experimental implementation of the spatial model," in *Proc. 6th ERCIM workshops*, Pehrson, B., Skärback, E. (eds.), Stockholm, June 1-3, 1994, pp. 53-71.

The MASSIVE-1 and the user-oriented evaluation of chapter 5 was published as:

> Greenhalgh, Chris, and Benford, Steve (1995), "Virtual Reality Tele-conferencing: Implementation and Experience", *Proc. Fourth European Conference on Computer Supported Cooperative Work (ECSCW'95)*, Stockholm, September, 10-14 September, 1995, Kluwer Academic Publishers, Dordrecht, pp. 165-180.

Extended with an earlier version of network analysis of chapter 6 (excluding audio and consideration of multicasting) this was then published as:

> Greenhalgh, Chris, and Benford, Steve (1995), "MASSIVE: A Virtual Reality System for Tele-conferencing", *ACM Transactions on Computer Human Interfaces (TOCHI)*, Volume 2, Number 3, pp. 239-261, ISSN 1073-0516, ACM Press, September 1995.

The concept of spatial trading (from chapter 6) in the context of MASSIVE-1 was published in:

> Greenhalgh, Chris, and Benford, Steve (1995), "MASSIVE: a Distributed Virtual Reality System Incorporating Spatial Trading", *Proc. 15th IEEE International Conference on Distributed Computing Systems (ICDCS'95)*, Vancouver, Canada, May 30-June 2, 1995, IEEE Computer Society, pp 27-34.

Details of the BT/JISC funded ITW trials of MASSIVE-1, including the network analysis of chapter 6 and the user modelling of appendix A have been published as:

> Greenhalgh, C, Bullock, A, Tromp, J, and Benford, S (1997), "Evaluating the network and usability characteristics of virtual reality tele-conferencing", *BT Technology Journal*, Vol. 15, No 4, October 1997.

The movement analysis component of the user modelling of appendix A (with extended consideration of the design implications) has been published as:

> Greenhalgh, C. (1997), "Analysing movement and world transitions in virtual reality tele-conferencing", in *Proc. Fifth European Conference on Computer Supported Cooperative Work (ECSCW'97)*, John A. Hughes, Wolfgang Prinz, Tom Rodden and Kjeld Schmidt (eds.), 1997, Kluwer Academic Publishers, pp. 313-328.

The third party object concept of chapter 7 has been published as:

> Benford, Steve, and Greenhalgh, Chris (1997), "Introducing Third Party Objects into the Spatial Model of Interaction", in *Proc. Fifth European Conference on Computer Supported Cooperative Work (ECSCW'97)*, John A. Hughes, Wolfgang Prinz, Tom Rodden and Kjeld Schmidt (eds.), 1997, Kluwer Academic Publishers.

The third party object concept focusing on the area of creating and representing crowds has been published as:

> Benford, S. D., Greenhalgh, C. M., Lloyd, D. L. (1997), "Crowded Collaborative Virtual Environments", in *Proc. 1997 ACM Conference*

on Human Factors in Computer Systems (CHI'97), Atlanta, Georgia, March 22-27, 1997, ACM Press.

The NOWninety6 "MASSIVE" real/virtual poetry performance using MASSIVE-2 (described in section 9.2.1) has been reported in:

Benford, Steve, Greenhalgh, Chris, Snowdon, Dave, and Bullock, Adrian(1997), "Staging a Public Poetry Performance in a Collaborative Virtual Environment", in *Proc. Fifth European Conference on Computer Supported Cooperative Work (ECSCW'97)*, John A. Hughes, Wolfgang Prinz, Tom Rodden and Kjeld Schmidt (eds.), 1997, Kluwer Academic Publishers, pp. 125-140.

The multicast-based network architecture of MASSIVE-2 has been presented in:

Greenhalgh, C., and Benford, S. (1997), "A Multicast Network Architecture for Large Scale Collaborative Virtual Environments", in Multimedia Applications, Services and Techniques - ECMAST'97, Proceedings Second European Conference, Serge Fdida and Michele Morganti (eds.), Milan, Italy, May 21-23, 1997, pp. 113-128, Springer.

B.2. Secondary Publications

MASSIVE-1 is used as an exemplar is the following papers:

Benford, S D. and Greenhalgh, C M. (1995), "A Spatial Approach to Speech and Gestural Control in Collaborative Virtual Environments", *Proc. Combined International Conference on Artificial Reality and Tele-Existence '95 and ACM Symposium on Virtual Reality Software and Technology' 95 (ICAT/VRST'95)*, Makuhari, Chiba, Japan, November 1995.

Benford, S. D., Bowers, J., Fahlén, L. E., Greenhalgh, C. M., Mariani, J. and Rodden, T. R. (1995), "Networked Virtual Reality and Co-operative Work", Presence: Teleoperators and Virtual Environments, Vol. 4, No. 4, Fall 1995, pp 364-386, MIT Press, ISSN1054-7460, Fall 1995.

Benford, Steve, Bowers, John, Fahlén, Lennart E., Greenhalgh, Chris, and Snowdon, Dave (1997), "Embodiments, avatars, clones and agents for multi-user, multi-sensory virtual worlds", *Multimedia Systems* (1997) 5: 93-104, Springer-Verlag.

Benford, Steve, Bowers, John, Fahlén, Lennart, and Greenhalgh, Chris, "Managing Mutual Awareness in Collaborative Virtual Environments," in Virtual Reality Software and Technology, Proceedings of the VRST'94 Conference, 23-26 August 1994, Singapore, Gurminder Singh, Steven K Feiner and Daniel Thalmann (eds.), pp. 223-236, Singapore: World Scientific Publishing.

The MASSIVE-2 version of WWW-3D is reported in:

Snowdon, Dave, Benford, Steve, Greenhalgh, Chris, Ingram, Rob, Brown, Chris, Lloyd, Dave, Fahlén, Lennart, and Stenius, Mårten (1997), "A 3D Collaborative Virtual Environment for Web Browsing" in *Virtual Reality Universe'97* April 2-5, 1997, Westin Santa Clara Hotel, California, USA.

B.3. Project Input

The work described in this thesis has contributed to the following projects.

- The EC ESPRIT III Basic Research Project COMIC (1992-1995). The original spatial model of interaction was developed within this project and the original development of MASSIVE-1 occurred within this context. MASSIVE-1 was used for international trials between partners in three countries over the Internet. This is the basis of the analysis of MASSIVE-1 reported in [Bowers, Pycock and O'Brien, 1996] and [Bowers, O'Brien and Pycock, 1996].
- The EPSRC funded Distributed Extensible Virtual Reality Laboratory project (DEVRL). MASSIVE-1 was also one of the test platforms within this project, with a stronger focus on logging and analysis.
- The BT/JISC funded Inhabiting the Web (ITW) project. The primary goal of this project was to hold a number of distributed meetings using MASSIVE-1 in order to gain experience with the technology and in particular to explore its networking requirements and possible approaches to measurements and analysis.
- The EPSRC large scale wide area real-time virtual reality project (a.k.a. HIVE, Huge Virtual Environments). The first year of this project included wide-area trials of MASSIVE-2 and Reading University's PaRADE system. This is informing the exploration of issues for future systems, in particular the combination of closely coupled interaction (e.g. with virtual objects) and awareness management.
- JTAP Virtual Campus project. This project is delivering CVE technologies to the UK postgraduate researcher community. MASSIVE-2 will be one of the tools used.

MASSIVE-2 is also making an important contribution to the following new (linked) projects.

- The EC ESPRIT I^3 research project eRENA.

- The directly funded BT project Network Architectures for Inhabited Television.
- The EPSRC Multimedia Networking for Inhabited Television project.

These projects are exploring the common theme of CVEs for arts, performance and entertainment. MASSIVE-2 is the core development platform for the last two projects which are focusing on the domestic/consumer application of Inhabited Television. MASSIVE-2 is also the platform for Nottingham University's involvement in the first project, which with the other partners involved spans a broader range of scenarios including artistic installations and performances.

References

Andrews, Gregory R. (1991), "Paradigms for process interaction in distributed programs", *ACM Computing Surveys*, Vol. 23, No. 1 (March 1991), pp. 49-90.

ANSA (1989), *The ANSA Reference Manual*. Architecture Projects Management Ltd., Poseidon House, Castle Park, Cambridge, UK.

Advanced Research Projects Agency (ARPA) (1994), "STOW 97 Program Plan", May.

Bajura, M., Fuchs, H., Ohbuchi, R. (1992), "Merging virtual objects with the real world: seeing ultrasound imagery within the patient", in *Conference proceedings of SIGGRAPH '92 on Computer graphics*, pages 203-210.

Barrus, J.W., Waters, R.C., and Anderson, D.B. (1996), "Locales: Supporting Large Multiuser Virtual Environments", in *IEEE Computer Graphics and Applications*, November 1996, Vol. 16, No. 6, pp. 50-57.

Bell, G., Parisi, A., and Pesce, M. (1995), The Virtual Reality Modelling Language, Version 1.0 Specification, URL: http://vag.vrml.org/vrml10c.html

Benford, S.D., and Fahlén, L.E. (1993), "A spatial model of interaction in large virtual environments", in *Proc. Third European Conference on Computer Supported Cooperative Work (ECSCW'93)*, Milano, Italy. Kluwer Academic Publishers, pp. 109-124.

Benford, S D. and Greenhalgh, C M. (1995), "A Spatial Approach to Speech and Gestural Control in Collaborative Virtual Environments", *Proc. Combined International Conference on Artificial Reality and Tele-Existence '95 and ACM Symposium on Virtual Reality Software and Technology' 95 (ICAT/VRST'95)*, Makuhari, Chiba, Japan, November 1995.

Benford, S., Bullock, A., Cook, N., Harvey, P., Ingram, R., and Lee, O. (1993), "From rooms to cyberspace: models of interaction in large virtual computer spaces", in *Interacting with Computers*, vol. 5, no. 2, pp. 217-237, Butterworth-Heineman Ltd.

Benford, Steve, Bowers, John, Fahlén, Lennart E., Mariani, John, and Rodden, Tom (1994), "Supporting Cooperative Work in Virtual Environments", *The Computer Journal*, Vol. 37, No. 8, 1994, pp. 653-668.

Benford, S. D., Bowers, J., Fahlén, L. E., Greenhalgh, C. M., Mariani, J. and Rodden, T. R. (1995), "Networked Virtual Reality and Co-operative Work", Presence: Teleoperators and Virtual Environments, Vol. 4, No. 4, Fall 1995, pp 364-386, MIT Press, ISSN1054-7460, Fall 1995.

Benford, S. D., Brown, C. C., Reynard, G. T. and Greenhalgh, C. M (1996), "Shared Spaces: Transportation, Artificiality and Spatiality", *Proc. ACM Conference on Computer Supported Cooperative Work (CSCW'96)*, Boston, pp. 77-86, ACM Press, Nov 16- 20, 1996.

Benford, S. D., Greenhalgh, C. M., Lloyd, D. L. (1997), "Crowded Collaborative Virtual Environments", in *Proc. 1997 ACM Conference on Human Factors in Computer Systems (CHI'97)*, Atlanta, Georgia, March 22-27, 1997, ACM Press.

Benford, Steve, Greenhalgh, Chris, Snowdon, Dave, and Bullock, Adrian(1997), "Staging a Public Poetry Performance in a Collaborative Virtual Environment", in *Proc. Fifth European Conference on Computer Supported Cooperative Work (ECSCW'97)*, John A. Hughes, Wolfgang Prinz, Tom Rodden and Kjeld Schmidt (eds.), 1997, Kluwer Academic Publishers, pp. 125-140.

Benford, Steve, Bowers, John, Fahlen, Lennart E., Greenhalgh, Chris, and Snowdon, Dave (1997), "Embodiments, avatars, clones and agents for multi-user, multi-sensory virtual worlds", *Multimedia Systems* (1997) 5: 93-104, Springer-Verlag.

Bowers, J. (1993), "Modelling Awareness and Interaction in Virtual Spaces", *Supplement to Proceedings of the 5th MultiG Workshop*, Stockholm-Kista, May 17, 1993, pp. S9-S24.

Bowers, John, Button, Graham, and Sharrock, Wes (1995), "Workflow from Within and Without: Technology and Cooperative Work on the Print Industry Shopfloor", *Proceedings of the Fourth European Conference on Computer Supported Cooperative Work (ECSCW'95)*, 10-14 September, 1995, Stockholm, Sweden, Hans Marmolin, Yngve Sundblad and Kjeld Schmidt (eds.), Kluwer Academic Publishers, Dordrecht, pp.51-66.

Bowers, John, Pycock, James, and O'Brien, Jon (1996), "Talk and Embodiment in Collaborative Virtual Environments", in *Proceedings of CHI'96*, Vancouver, Canada, New York: ACM Press, 1996.

Bowers, John, O'Brien, Jon, and Pycock, James (1996), "Practically Accomplishing Immersion: Cooperation in and for Virtual Environments", in *Proceedings of CSCW'96*, New York: ACM Press, 1996.

Braden, R. , Zhang, L., Berson, S., Herzog, S., and Jamin, S. (1996), "Resource ReSerVation Protocol (RSVP) = Version 1 Functional Specification", Internet draft <draft-ietf-rsvp-spec-14.ps>, October 29, 1996.

Brand, S. (1987), *The Medialab - Inventing the future at MIT*, Viking Penguin, 1987, ISBN 0-670-81442-3, p. 91-93.

Bricken, William, and Coco, Geoffrey (1994), "The VEOS Project", *Presence*, Vol. 3, No. 2, Spring 1994, 111-129.

Broll, W. (1997), "Distributed Virtual Reality for Everyone -a Framework for Networked VR on the Internet", in *Proceedings of the IEEE Virtual Reality Annual International Symposium 1997* (VRAIS'97), March 1-5, Albuquerque, NM.

Brown, Nat, and Kindel, Charlie (1996),
INTERNET-DRAFT <draft-brown-dcom-v1-spec-01.txt>, November, 1996.
http://www.microsoft.com/oledev/olecom/draft-brown-dcom-v1-spec-01.txt

Carlsson, C. and Hagsand, O. (1993), "DIVE - A Multi-User Virtual Reality System", in *Proc. VRAIS'93, IEEE Virtual Reality Annual International Symposium*, pp. 394-400.

Chapanis, Alphonse (1975), "Interactive Human Communication", *Scientific American*, 232 (3): 36-42, 1975, reproduced in *Computer Supported Cooperative Work: A Book of Readings*, Ed. Irene Grief, pp. 127-140, Morgan Kaufman, San Mateo, CA.

Curtis, Pavel, and Nichols, David (1994), "MUDs Grow Up: Social Virtual Reality in the Real World", *Proc. IEEE Computer Conference*, January, 1994.

Das, Tapas K. , Singh, Gurminder, Mitchell, Alex, Kumar, P Senthil, McGee, Kevin (1997), "NetEffect: A Network Architecture for Large-Scale Multi-user Virtual World", in *Proc. IEEE Sixth Workshops of Enabling Technologies: Infrastructure for Collaborative Enterprises*, June 18-20, 1997, MIT, Cambridge, MA.

Deering, Stephen (1989), "Host Extensions for IP Multicasting", IETF RFC 1112 (Aug. 1989).

Deering, Stephen E., and Cheriton, David R. (1990), "Multicast Routing in Datagram Internetworks and Extended LANs", *ACM Transactions on Computer Systems*, Vol. 8, No. 2, May

1990, Pages 85-110.

Division Ltd. (1996) *dVS Developers Guide. Rev. 3.0.* Division Ltd., Bristol, UK. (http://www.division.co.uk/)

Ellis, C.A, Gibbs, S.J., and Rein, G.L. (1990), "Design and use of a group editor". In *Engineering for Human Computer Interaction.* G. Cockton, Ed., North-Holland, Amsterdam, 1990, 13-25.

Ellis, C.A. , Gibbs, S.J., and Rein, G.L. (1991), "Groupware: Some Issues and Experiences", *Communications of the ACM,* January 1991, Vol. 34, No. 1.

Flores, Fernando, Michael Graves, Brad Hartfield and Terry Winograd (1988), " Computer Systems and the Design of Organisational Interaction". *ACM Transactions on Office Information Systems* 6(2): 153-172.

Fahlén, Lennart E., Brown, Charles Grant, Stahl, Olov, and Carlsson, Christer (1993), "A Space Based Model for User Interaction in Shared Synthetic Environments", in *Proc. INTERCHI'93,* 24-29 April 1993, Amsterdam. pp. 43-48. ACM Press.

Foucault, M. (1977) *Discipline and Punish.* Harmondsworth: Penguin.

Funkhouser, T.A. (1995), "RING: A Client-Server System for Multi-User Virtual Environments." *Computer Graphics, 1995 SIGGRAPH Symposium on Interactive 3D Graphics,* Monterey, CA., April 1995, pp. 85-92.

Funkhouser, T.A. (1996), "Network Topologies for Scalable Multi-User Virtual Environments", in *Proc. 1996 IEEE Virtual Reality Annual International Symposium* (VRAIS96), San Jose, CA, April, 1996.

Funkhouser, T.A., Séquin, C.H., and Teller, S.J. (1992), "Management of large amounts of data in interactive building walk-throughs", *ACM SIGGRAPH Special Issue on 1992 Symposium on Interactive 3D Graphics,* pp. 11-20.

Gale, S. (1991), "Adding audio and video to an office environment" in Bowers, J., and Benford, S. (eds) *Studies in Computer Supported Cooperative Work: Theory, Practice and Design,* Elsevier.

Giddens, A. (1984). *The Constitution of Society.* Polity, Cambridge.

Glance, Natalie S., Pagani, Daniele S., and Pareschi, Remo (1996), "Generalized Process Structure Grammars (GPSG) for Flexible Representations of Work", in *Proc. Computer Supported Cooperative Work (CSCW) 96,* Cambridge MA USA, pp. 180-189.

Greenhalgh, C. (1994), "An experimental implementation of the spatial model," in *Proc. 6th ERCIM workshops,* Pehrson, B., Skarback, E. (eds.), Stockholm, June 1-3, 1994, pp. 53-71.

Greenhalgh, C. M. (1996), "MASSIVE: evolution and reflection", 3rd UK Virtual Reality Special Interest Group (VRSIG) workshop, July 1996, DeMontfort University, Leicester.

Greenhalgh, C. (1997), "Analysing movement and world transitions in virtual reality tele-conferencing", in *Proc. Fifth European Conference on Computer Supported Cooperative Work (ECSCW'97),* John A. Hughes, Wolfgang Prinz, Tom Rodden and Kjeld Schmidt (eds.), 1997, Kluwer Academic Publishers, pp. 313-328.

Greenhalgh, Chris, and Benford, Steve (1995a), "MASSIVE: A Virtual Reality System for Tele-conferencing", *ACM Transactions on Computer Human Interfaces (TOCHI),* Volume 2, Number 3, pp. 239-261, ISSN 1073-0516, ACM Press, September 1995.

Greenhalgh, Chris, and Benford, Steve (1995b), "Virtual Reality Tele-conferencing: Implementation and Experience", *Proc. Fourth European Conference on Computer Supported Cooperative Work (ECSCW'95),* Stockholm, September, 10-14 September, 1995, Kluwer Academic Publishers, Dordrecht, pp. 165-180.

Greenhalgh, Chris, and Benford, Steve (1995c), "MASSIVE: a Distributed Virtual Reality

System Incorporating Spatial Trading", *Proc. 15th IEEE International Conference on Distributed Computing Systems (ICDCS'95)*, Vancouver, Canada, May 30-June 2, 1995, IEEE Computer Society, pp 27-34.

Greenhalgh, Chris, and Benford, Steve (1997), "Boundaries, Awareness and Interaction in Collaborative Virtual Environments", Published in the *proceedings of the 6th International Workshop on Enabling Technologies: Infrastructure for Collaborative Enterprises (WET-ICE)*, June 18-20, 1997, Cambridge, Massachusetts, USA.

Greenhalgh, C, Bullock, A, Tromp, J, and Benford, S (1997), "Evaluating the network and usability characteristics of virtual reality tele-conferencing", *BT Technology Journal*, Vol 15, No 4, October 1997.

Greif, Irene (1988). *Computer-Supported Cooperative Work: A Book of Readings*. San Mateo, California: Morgan Kaufmann Publishers.

Grimsdale, C. (1991), "dVS - Distributed Virtual Environment System". In *Proc. Computer Graphics '91*. London, ISBN 0 86353 282 9.

Hagsand, Olof (1996), "Interactive Multiuser VEs in the DIVE System", *IEEE Multimedia*, Spring 1996, Vol. 3, No. 1, pp. 30-39, IEEE Computer Society, ISSN 1070-986X

Hardman, Vicky, Sasse, Angela, Handley, Mark, and Watson, Anna (1995), "Reliable Audio for Use over the Internet", in *Proceedings of Inet'95*, June 1995, Honolulu, Hawaii.

Heath, C., and Luff, P. (1991) "Collaborative activity and technological design: task coordination in London Underground control rooms". In Bannon, L., Robinson, M., and Schmidt, K. (eds), *Proc. ECSCW'91*. Kluwer, Dordrecht.

Holbrook, Hugh W., Singhal, Sandeep K., and Cheriton, David R. (1995), "Log-Based Receiver-Reliable Multicast for Distributed Interaction Simulation", In *Proceedings of SIGCOMM '95*, Cambridge, MA, August, 1995. ACM SIGCOMM.

Hughes, J.A., Randall, D., and Shapiro, D. (1992), "Faltering from ethnography to design". In *Proc. CSCW'92*, Toronto.

IEEE, (1993): Institute of Electrical and Electronic Engineers, International Standard, ANSI/ IEEE Std 1278-1993, Standard for Information Technology, Protocols for Distributed Interactive Simulation (March 1993).

ISO (1990): ISO, Information Processing Systems - Open Systems Interconnect - Open Distributed Processing (ODP), ISO/TC 97/SC21/WG7

ITU-T Recommendation X.901 (1995) | ISO/IEC 10746-1:1995, Open Distributed Processing - Reference Model - Overview.

Jacobson, V. (1992), VAT manual pages, Lawrence Berkeley Laboratory (LBL) February 1992.

Johansen, Robert, and Bullen, Christine (1984), "Thinking Ahead: What to Expect from Teleconferencing", *Harvard Business Review* (Mar.-Apr. 1984), pp. 4-10, reproduced in *Computer Supported Cooperative Work: A Book of Readings*, Ed. Irene Grief, Morgan Kaufman, San Mateo, CA., pp. 185-198.

Johnston, R.S. (1987), "The SimNet Visual System", *Proc. 9th Inter-Service/Industry Training Systems and Education Conf.*, Nov. 1987, pp. 264-273.

Kazman, R. (1993), "Making WAVES: on the Design of Architectures for Low-end Distributed Virtual Environments", *Proc. IEEE Virtual Reality Annual International Symposium (VRAIS'93)*, Washington, USA.

Kazman, Rick (1995), "Load Balancing, Latency Management and Separation of Concerns in a Distributed Virtual World", in *Parallel Computations - Paradigms and Applications*. A. Zomaya (ed.) Chapman and Hall, 1995, pp. 480-497.

Lea, Rodger, Honda, Yasuaki, and Matsuda, Kouichi (1997), "Virtual Society: Collaboration in

3D Spaces on the Internet", *Computer Supported Cooperative Work: The Journal of Collaborative Computing*, 6:227-250, 1997. Kluwer Academic Publishers, Netherlands.

Lefebvre, H. (1990) *The Politics of Space*. Cambridge: Polity.

Leland, M.D.P., Fish, R.S., and Kraut, R.E. (1988), "Collaborative document production using Quilt". In *Proceedings of the Conference on Computer Supported Cooperative Work* (Portland, Oreg., Sept. 26-28). ACM New York, 1988, pp. 206-215.

Macedonia, Michael R., and Brutzman, Donald P. (1994), "MBone Provides Audio and Video Across the Internet", *IEEE Computer*, April 1994, pp. 30-36.

Macedonia, Michael R., Zyda, Michael J., Pratt, David R., Barham, Paul T., and Zeswitz, Steven (1994), "NPSNET: A Network Software Architecture for Large-Scale Virtual Environments", *PRESENCE*, Vol. 3, No. 4, Fall 1994, pp. 265-287.

Macedonia, Michael R., Zyda, Michael J., Pratt, David R., Brutzman, Donald P., and Barham, Paul T. (1995), "Exploiting Reality with Multicast Groups: A Network Architecture for Large-scale Virtual Environments", in *Proc. 1995 IEEE Virtual Reality Annual International Symposium (VRAIS'95)*, 11-15 March, 1995, RTP, North Carolina.

Magee, Jeff, Dulay, Naranker, and Kramer, Jeff (1994), "A Constructive Development Environment for Parallel and Distributed Programs", in *IEE/IOP/BCS Distributed Systems Engineering*, 1(5): 304-312, Sept. 1994.

Marx, K. (1857) *Capital*, volume 1. Moscow: Progress.

Milgram, Paul, and Kishino, Fumio (1994), "A Taxonomy of Mixed Reality Virtual Displays", IEICE Transations on Information Systems, Vol. E77-D, No. 12, December 1994.

Morningstar, Chip, and Farmer, F.Randall (1991), "The Lessons of Lucasfilm's Habitat", in *Cyberspace: First Steps*, edited by Michael Benedict, MIT Press, Cambridge, MA., pp. 273-301, 1991.

Nakanishi, Hideyuki, Yoshida, Chikara, Nishimura, Toshikazu, and Ishida, Toru (1996), "FreeWalk: Supporting Casual Meetings in a Network", in *Proc. Computer Supported Cooperative Work '96 (CSCW'96)*, Cambridge, MA., USA, ACM.

Ohya, J., Kitamura, Y., Takemura, H., Kishino, F., Terashima, N. (1993), "Real-time Reproduction of 3D human Images in Virtual Space Teleconferencing", *Proc. VRAIS'93*, IEEE, Seattle, Washington, September 1993, pp. 408-414.

Robinson, Mike (1993), "Design for unanticipated use.....", in *Proceedings of the Third European Conference on Computer-Supported Cooperative Work*, 13-17 September, 1993, Milan, Italy. G De Michelis, C. Simone and K. Schmidt (Editors). pp. 187-202. Dordrecht: Kluwer Academic Publishers.

Rodden, Tom (1996), "Populating the Application: A Model of Awareness for Cooperative Applications", in *Proc. ACM 1996 Conference on Computer Supported Cooperative Work (CSCW'96)*, November 16-20, 1996, Boston, Massachusetts, USA, ACM Press, pp. 87-96.

Rodden, T., Mariani, J.A., and Blair, G. (1992), "Supporting Cooperative Applications", *Computer Supported Cooperative Work (CSCW): An International Journal*, 1992.

Root, R.W. (1988), "Design of a Multi-Media Vehicle for Social Browsing", In *Proc. CSCW'88*, Portland, Oregon, September 26-28, 1988, pp. 25-38.

Salmony, Michael (1995), "Interactive TV as seen from the information processing industry", in Proc. Imagina, Monte-Carlo, 1-3 February, 1995, pp. 142-155, ISBN 2-86938-113-1.

Sandor, Ovidiu, Bogdan, Cristian, and Bowers, John (1997), "Aether: An Awareness Engine for CSCW", in *Proc. Fifth European Conference on Computer Supported Cooperative Work (ECSCW'97)*, John A. Hughes, Wolfgang Prinz, Tom Rodden and Kjeld Schmidt (eds.), 1997, Kluwer Academic Publishers.

Sarin, Sunil, and Greif, Irene (1985), "Computer-Based Real-Time Conferencing Systems",

IEEE Computer, Vol. 18, No. 10, October 1985, pp. 33-45.

Schmidt, Kjeld, and Bannon, Liam (1992), "Taking CSCW Seriously. Supporting Articulation Work", *Computer Supported Cooperative Work (CSCW)*, Vol. 1, Nos. 1-2, 1992, Kluwer Academic Publishers, Dordrecht.

Shaw, Chris, Liang, Jiandong, Green, Mark, and Sun, Yunqi (1992), "The Decoupled Simulation Model for Virtual Reality Systems", in *Proc. ACM CHI'92*, May 3-7, 1992, pp. 321-328.

Singhal, Sandeep Kishan (1996), "Effective Remote Modelling in Large-Scale Distributed Simulation of Visualization Environments", PhD thesis, Stanford University, Department of Computer Science, August 1996.

Singhal, S.K., and Cheriton, D.R. (1996), "Using Projection Aggregations to Support Scalability in Distributed Simulation", in *Proc. 1996 International Conference on Distributed Computing Systems* (ICDCS'96), 1996.

Singh, Gurminder, Serra, Luis, Pang, Willie, and Ng, Hern (1994), "BrickNet: A Software Toolkit for Network-Based Virtual Worlds", *Presence*, Vol. 3, No 1, Winter 1994, 19-34.

Slater, M., Usoh, M., Benford, S., Snowdon, D., Brown, C., Rodden, T., Smith, G., Wilbur, S. (1996), "Distributed Extensible Virtual Reality Laboratory (DEVRL) - A Project for Cooperation in Multi Participant Environments", in 3rd Eurographics Workshop on Virtual Environments, M. Goebel (ed.), 19-20 February 1996, Monte Carlo, ISSN 1024-0861.

Snowdon, David N., and West, Adrian J. (1994), "AVIARY: Design issues for future large-scale Virtual Environments", *Presence* 3(4), 1994.

Snowdon, D. N., Greenhalgh, C. M. and Benford, S. D. (1995), "What You See is Not What I See: Subjectivity in Virtual Environments", *Proc. Framework for Immersive Virtual Environments - FIVE'95*, London, 18-19 December, 1995.

Snowdon, Dave, Benford, Steve, Greenhalgh, Chris, Ingram, Rob, Brown, Chris, Lloyd, Dave, Fahlén, Lennart, and Stenius, Mårten (1997), "A 3D Collaborative Virtual Environment for Web Browsing" in *Virtual Reality Universe'97* April 2-5, 1997, Westin Santa Clara Hotel, California, USA.

Stefik, M., Bobrow, D.G., Foster, G., Lanning, S., and Tatar, D. (1987), "WYSIWIS Revised: Early Experiences with Multi-User Interfaces", *ACM Transactions on Office Information Systems*, Vol. 5, No. 2, April, 1987, pp. 147-167.

Thalmann, D. (1993), "Using Virtual Reality Techniques in the Animation Process", in *Virtual Reality Systems*, Earnshaw, R.A., Gigante, M.A., and Jones, H. (eds.), Academic Press, 1993, ISBN 0-12-227748-1.

Tromp, J. G., (1995a). "An Evaluation of MASSIVE by Measuring Telepresence and Immersion", COMIC Deliverable, ESPRIT Basic Research Project 622S.

Tromp, J. G., (1995b). "Virtual Reality in the Office: Evaluation of MASSIVE by Measuring Telepresence and Immersion", Poster at European Computer Supported Cooperative Work, September '95, Stockholm, in: Supplement Proceedings ECSCW'95.

Vinoski, Steve (1997), "CORBA overview: CORBA: Integrating Diverse Applications Within Distributed Heterogeneous Environments", *IEEE Communications Magazine*, Vol. 14, No. 2, February, 1997.

Wang, Qunjie, Green, Mark, and Shaw, Chris (1995), "EM - An Environment Manager for Building Networked Virtual Environments", in *Proc. IEEE Virtual Reality Annual International Symposium 1995*.

Weiser, Mark (1993), "Some Computer Science Issues in Ubiquitous Computing", Communications of the ACM, July 1993, Vol. 36, No. 7, pp. 74-84.

Index

MIX
Papier aus verantwortungsvollen Quellen
Paper from responsible sources
FSC® C105338

If you have any concerns about our products,
you can contact us on
ProductSafety@springernature.com

In case Publisher is established outside the EU,
the EU authorized representative is:
Springer Nature Customer Service Center GmbH
Europaplatz 3, 69115 Heidelberg, Germany

Printed by Libri Plureos GmbH
in Hamburg, Germany